Essential **Boat Radar**

Bill Johnson

Essential
Boat Radar

Bill Johnson

WILEY ✶ NAUTICAL

This edition first published 2009
© 2009 Bill Johnson

Registered office
John Wiley & Sons Ltd, The Atrium, Southern Gate, Chichester, West Sussex, PO19 8SQ, United Kingdom

For details of our global editorial offices, for customer services and for information about how to apply for permission to reuse the copyright material in this book please see our website at www.wiley.com.

Library of Congress Cataloging-in-Publication Data
Johnson, Bill.
Essential boat radar / Bill Johnson.
p. cm.
Includes bibliographical references and index.
ISBN 978-0-470-77811-1 (pbk. : alk. paper)
1. Radar in navigation. 2. Boats and boating--Radar equipment. I. Title.
VK560.J64 2009
623.89'33--dc22
2008052052

A catalogue record for this book is available from the British Library.

Set in Humnst777EU by Laserwords Madison, Madison WI, USA

Printed and bound by SNP Leefung Printers Ltd, China

CONTENTS

C

C

1 Introduction

"Does anyone know how the radar works?"

Radar has become much more accessible for small craft in recent years. This is chiefly because the display has become smaller and flatter, and far easier to install on a small yacht or motor boat. The same technological advances have also meant that it consumes less power (a major consideration in sailing boats), and can be usefully integrated with other devices, such as GPS and chart plotters (it can even be built into the same display). It's also quite a bit cheaper than it was (and anyway, people seem to have more money to spend on electronic gadgets!).

Radar really does have an element of magic about it. This is because it can 'see' things that cannot be seen by eye – in conditions when very little, or nothing, *can* be seen by eye – and determine a few very useful things about them, such as their exact distance and direction from the boat. On the other hand, what it tells you has limitations; and it is important to understand those limitations – and the reasons behind them – in order to use radar effectively. And with a radar system (no less than with other devices such as chart plotters), you need to know how to control the equipment in order to get a useable picture. Only a bit of learning can enable you to do this.

Most people sense this as soon as they look at a radar display. With a chart plotter, the designers have made it as obvious as possible for the user (or casual viewer) to interpret what is displayed to them on the screen, and in any case it is pretty easy to interpret as long as you have seen a chart or map before. With radar, you get a picture, sure – but it is rather different to any picture that you normally have to deal with. Although you can generally see what is going on, it is basically a slightly confusing pattern of blobs. Most people would definitely think 'I need to understand what is going on here', and it is for them that this book is written.

The recent technological advances have mainly been in the part of the device which processes and displays the picture. This has benefited hugely from modern microprocessing, which enables you to do almost anything with the picture that the radar detects: move it, turn it around, zoom in to areas of interest, etc. The basic technology for getting the picture, however, has not really changed very much; and it is this underlying process that you need to understand – a bit – in order to understand the picture you are getting. At the same time, I do appreciate that, generally, people reading this book will prefer to go boating than to spend ages learning about the finer points of microwave physics (or, indeed, staring at a radar screen).

1

So in this book I try to take a fairly direct approach to the task of getting you into a position where you can use radar. First, I explain what radar is and how it works (which isn't very different from when it was invented in its current form in the early 1940s). Then, in Chapter 3, I explain how to operate the machine and controls to get a good picture. At this stage I deliberately ignore the more advanced and exciting functionality of modern radar, because I want to get you to a point where you can use it as quickly as possible.

Having got a picture, what do you use it for? Well, Chapters 4 and 5 tackle the practical approach to two things that the radar can help you with: collision avoidance (Chapter 4), and navigation and pilotage (Chapter 5). This, of course, is the exciting bit. Radar can give you a lot more confidence, particularly when the visibility is poor or things are confusing at night. Radar is particularly useful for less experienced mariners who might want to confirm or quantify something, for example about a ship that they can just see in the distance. Radar can tell you how big it is, and how far away.

I return to the physics in Chapter 6, and explain some of the confusion and anomalies that you can get with the picture. These are pretty much commonsense once you know about them – what happens when the radar signal is reflected twice, etc – but they are the sort of effects that you are going to get occasionally, and need to be confident in interpreting.

Chapter 7 considers some of the more advanced functions that modern radars provide, and when they might be useful. What this book can't do, of course, is explain the controls and user interface of any particular radar system, because they are all different. But after reading this chapter you will know what you are looking for when you start reading the *Owner's Handbook* for your own radar.

Chapter 8 is a very brief 'how to' guide for choosing and installing a radar system – principally what to look for in the performance and features of different makes and models, and the issues that need to be considered for installation.

A glossary/index, and further references are also included.

I hope that you will get a great deal of satisfaction in overcoming the initial challenges of using radar, and in building up a new and very useful skill. To begin with, you will not have anywhere near the skills and abilities of a trained radar operator, and you won't want to rely on this device excessively (it would be mad to go out in thick fog just to see how you get on!). After a while, you will probably find that this is a very useful device for backing up other observations; and if you use it regularly you will:

■ become familiar with the control interface on your own radar; and

■ be able to place more confidence in your radar observations for collision avoidance and pilotage.

Also, you will learn more about what radar can do, and what its limitations are. This is a very useful lesson in itself, because most of the large ships we encounter use radar a great deal for detecting and avoiding small craft like our own. Learning what is involved, and seeing how well our own radar performs in different conditions, is an important lesson. At the very least, it will probably make you go out and buy a better radar reflector for your own boat!

Good luck and good boating!

2 Radar – what it does, and how it works

How radar works

Radar transmits **microwaves** in a narrow **beam**, and then detects the returning echoes of those waves from objects in their path.

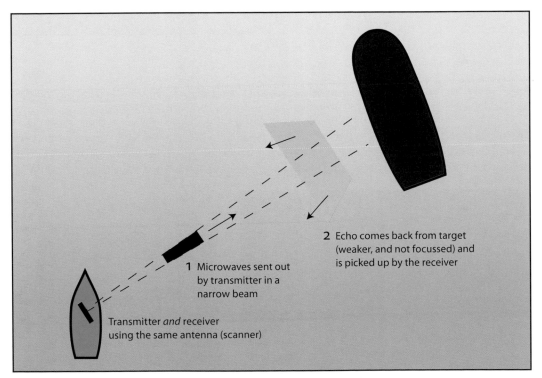

1 Microwaves sent out by transmitter in a narrow beam

2 Echo comes back from target (weaker, and not focussed) and is picked up by the receiver

Transmitter *and* receiver using the same antenna (scanner)

Figure 2.1 *Transmitted microwaves, and the returning echo from a ship*

First, a brief explanation about microwaves. The spectrum of electromagnetic waves extends from radio waves at low frequencies to visible light at much higher frequencies (and also beyond these, in both directions). Between the radio frequencies and light frequencies lie microwaves: they are officially designated as Extremely High Frequency (EHF) radio waves. As well as radar, microwaves are used for microwave ovens and communication (line of sight between those big towers you

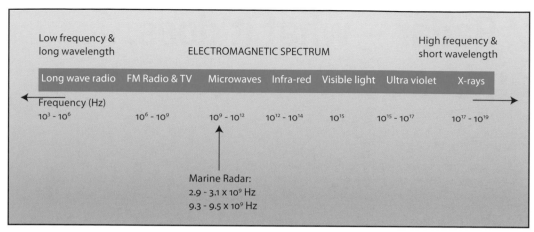

Figure 2.2 *The electromagnetic spectrum from radio to x-rays*

see on hills). In the sort of radar we are concerned with, this narrow beam is swept round in a horizontal circle, so that **echo returns** (also known as **targets**, betraying radar's military origins) can be detected all round the boat. These returns are then shown on a display.

The picture on the display is essentially a plan view from directly above your vessel and the surrounding targets, like a chart. It gives a representation of the *angle* and *range* of the targets detected. In its basic version, the boat's radar is at the centre, the boat's heading is straight up

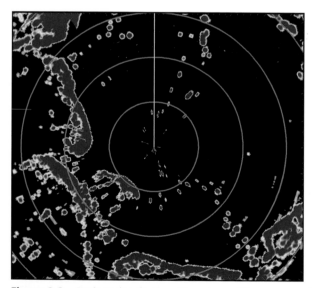

Figure 2.3 *Basic radar display. Plan view with own vessel at the centre*

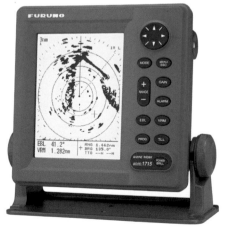

Figure 2.4 *Basic radar display on small monochrome system*

the screen at the 12 o'clock position, and you get a circular picture of the targets around you. This kind of picture is sometimes known by the technical term **plan position indicator**, or **PPI**.

Understanding what a radar does

When you look at a chart plotter, you are shown a great deal of information about the surroundings: coastlines, shape of the nearby land and, of course, the depth of the surrounding water. All of this comes from stored data – someone else's knowledge about the geography of the area – plus just one other important piece of information that your vessel has 'sensed': its own position, derived from satellite radio signals by the GPS receiver on board.

The radar picture is totally different. Everything in the displayed picture is actually sensed by the beam sweeping round the boat. (I exclude symbols, lines and other data which are added in by the processor: marker lines, data boxes, chart overlays, etc.) This means that, in one important respect, this picture is more useful than the chart plotter's: this is the real thing, detected now, rather than being information from a database. On the other hand, the overall picture is less obvious to interpret, because only the radar-detected information is presented, and there are limitations both to what can be detected and to what you can know about an object detected in this way.

This is why you have to understand the radar sensor – what it can and can't do – in order to interpret the picture.

There is a great temptation to lose patience with this. For example, you know there is an island over there. The chart plotter will show its exact shape all round (it will even tell you its name!) and anyway you will recognize it easily on the display from its shape and your knowledge of the area.

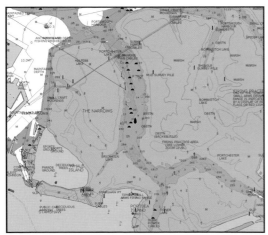

Figure 2.5 *Chart plotter display*

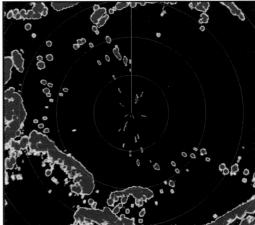

Figure 2.6 *Corresponding radar picture*

The radar, however, gives you a confusing smudge representing the edge of the island and nothing else – certainly not the coastline on the remote side of the island, and so not its overall shape. Worse than that, there are gaps in the smudge and its edges do not even coincide with the charted coastline. Well, this is a waste of time, isn't it? Just use the chart plotter.

But remember the power of really sensing something that's actually there. In this respect, it's just like seeing it. A blurred and indistinct visual sight of the coast can be confusing, but we trust it more (and rightly so) than the chart. In the same way, the radar 'sees' what's really there, today: including, for example, other vessels and things that the chart maker didn't know about.

How radar 'sees'

As previously explained, radar uses microwaves; but 'seeing' with radar is analogous to sweeping a strong searchlight around us in the dark and recording when it detects something. In some respects it is better than that: it can 'see' through fog, not just the dark, and it can also work out the exact distance of the object detected. Its range is also considerably greater than most searchlights.

On the downside, radar can't see the detected object's detail or shape with any great precision. Anything in the beam of the 'searchlight' is recorded as a target (a blob on the display) whilst it is in the beam; and because the beam is relatively wide at its outer end, more distant targets are also quite wide on the display. The advantage that we have when actually looking at some object picked out by a real searchlight (i.e. seeing its colour and detail, which our eyes do by focusing and analyzing the light that's returned), is not available to radar. And exactly as with the searchlight, radar can't see over the horizon, and can't see anything that's hidden behind something else. Radar sees the 'face' of the things it picks out as returns – just the surface that's turned towards it.

Grasp those facts and you're almost there.

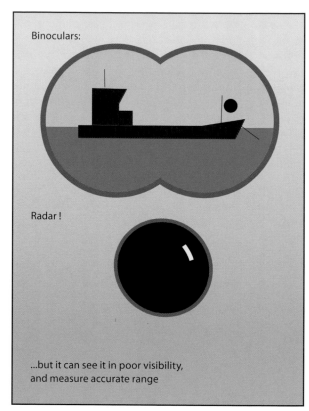

Figure 2.7 *Comparison between radar picture and binoculars*

Getting technical

There are limitations to the analogy of a searchlight beam, useful though it is. The actual technology of a radar beam is really quite simple to understand (if not to engineer).

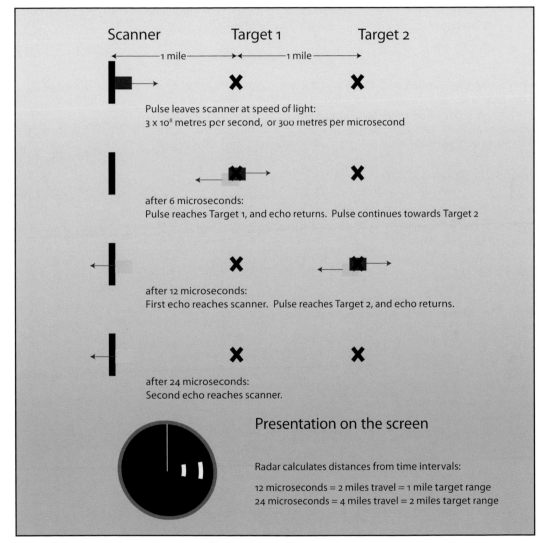

Figure 2.8 *How radar measures and displays target distance*

Measuring distance

You might wonder how radar actually measures the distance to the target. What it does is use **pulses** of microwave radiation rather than a continuous beam. It then times the interval between sending a pulse out and detecting the reflection of that pulse from the object – rather as you might time the interval between a shout and its echo off a cliff. From this time interval, the distance of the returning object can be determined by a simple formula:

distance = speed x time

where *distance* travelled by the reflected pulse (i.e. twice the range of the object) = *speed of light* x the *time* interval recorded (from send to receive).

(Interestingly, GPS also uses the speed of light to calculate distance from each satellite, and thereby determine our position.)

Measuring direction

The radar beam (as a rapid series of pulses) is swept around in a circle just as in our searchlight analogy. The radar knows the direction of a return, because it knows where the scanner is pointing when it receives the return: relative, of course, to the platform (the boat) that the scanner is attached to.

Basic radar picture display

In the basic radar display, targets are shown at the same angle from the vertical ship's heading line that the scanner is pointing in when they are detected. The picture is, in fact, painted in a line sweeping round the screen like a clock hand, exactly as the scanner itself sweeps round. The targets are shown on the display at a distance from the centre corresponding to the target range, so you get a 'view from above' of the targets around you, with your own boat at the centre.

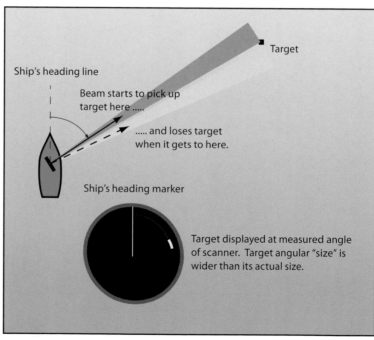

Figure 2.9 *Detecting a target with a radar beam. If the object is small, displayed target will be the same 'width' as the beam.*

Beam width

However, the **beam** has a finite angular **width**. Anything in the beam is recorded as a return for the whole time it is in the beam, so a narrow target is spread in the direction of the circular sweep, showing a wider return on the screen than the actual width of the object. Also, the radar will not be able to distinguish between two objects at the same range which are in the beam at the same time. And when it comes to measuring the bearing of the target, the precision of this measurement depends on the width of the beam.

So it is desirable to have a beam which is narrow in the horizontal direction. It is, however, much wider in the vertical direction so that targets can be detected quite close to the boat, and they will be detected even when the boat is rolling.

2

Picture stabilization

The measurement and display of direction works well if the boat is absolutely steady, but can be less than perfect on a small boat in a rough sea, when its heading is changing quite a lot. (After all, it's not very easy to use binoculars or take bearings in that situation, either.) In this case, the basic radar picture will twist back and forth as the boat does (and in the opposite direction): the return on the first sweep is placed at position A, and on the next sweep the same object will seem to the radar as if it is in position B. To say the least, this can be extremely confusing when using and interpreting the picture.

There is a solution to this problem, but it means departing from the basic radar display where the ship's heading is vertically up the screen. Most modern radars can process additional information from other sources and change the display accordingly. (If you have an old fashioned radar this is not possible, and you can stop reading this section now!)

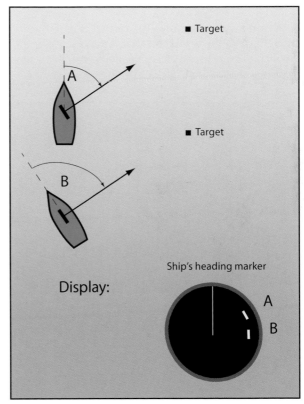

Figure 2.10 *Unstabilized display. The target will be displayed in different positions as the boat yaws.*

If you provide the radar processor with information on precisely where the boat is pointing from one fraction of a second to the next, then it can work out the *absolute* direction that the scanner is pointing in, and use that to display targets instead of using the simple measured direction relative to the boat. This is called an **azimuth stabilized** display, and it does not suffer from the yawing motion of the boat in the same way that the basic display does. There's more on this in Chapter 7 but it's mentioned here because it can be a particularly helpful facility, making the radar easier to use.

In practice, two types of stabilized display are generally available:

- **north-up**, where — as on a chart — north is vertically up on the screen;

- **course-up**, where the boat's intended heading (as opposed to its actual heading) is vertically up.

Both give a steady picture. The former is easier to relate to the chart for navigational purposes. The latter has the attraction that it looks just like an improved version of the basic display, and the picture corresponds (very nearly) to what you see around the boat: e.g. an object spotted 60° to port is shown at 10 o'clock on the display.

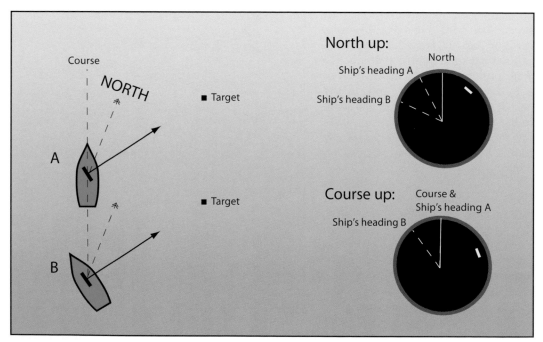

Figure 2.11   *Same situation with stabilized displays: north-up and course-up. The target will be displayed in the same position as the boat yaws.*

Where can radar see?

The microwaves used by radar behave similarly to light (or their other neighbour on the electromagnetic spectrum, VHF radio waves) when it comes to going round corners – i.e. they don't, except in peculiar atmospheric circumstances. They go in pretty well straight lines. This fact enables us to get a useful picture, but remember, the radar can *only* detect objects in its 'line of sight'. Our searchlight analogy comes in useful again here: radar, too, can only see in a straight line, so *cannot* detect objects below its antenna's horizon. Nor can it detect anything that is completely hidden behind another object (the exception to this is something like a heavy rain shower: although pretty effective at scattering visible light, it does not stop or reflect all the microwaves, so radar can 'see' through it).

The height of the antenna (like the height of a VHF radio antenna) determines how far it can see to its horizon. The formula is:

horizon distance *(in nautical miles)* = **2.21 x** √**antenna height** *(in metres)*

[maths reminder: the √ symbol means 'square root', e.g. √4 = 2, √9 = 3, √100 = 10]

This is the distance at which it can see targets at sea level. It is useful to be aware of the horizon distance for your particular radar installation, because you really want to detect things at sea level: small vessels and low coastlines. Beyond this distance, you will not detect them on radar.

But you will see other things which stick up above the horizon: larger vessels, cliffs and higher land, perhaps inland from the low coastline. The full formula is:

radar range *(in nautical miles)* = **2.21 x (**√**antenna height +** √**target height)** *(in metres)*

2

2

Examples of radar ranges

Own antenna height	Typical target		Radar range (nautical miles)
Small motor boat:			
3m	Small vessel/buoy (or *horizon range*)		3.8
	Large ship	20m	13.7
	Land at St Catherine's Pt	200m	35.1
Small yacht mast:			
6m	Small vessel/buoy (or *horizon range*)		5.4
	Large ship	20m	15.3
	Land at St Catherine's Pt	200m	36.7
Large yacht mast:			
12m	Small vessel/buoy (or *horizon range*)		7.7
	Large ship	20m	17.5
	Land at St Catherine's Pt	200m	38.9

So at a distance, hills and mountains will be seen, but they will look like islands even when they are, in reality, joined together – and the returns will be from inland, not from the coast. This is something to bear in mind if you want to use the radar to measure distance from land.

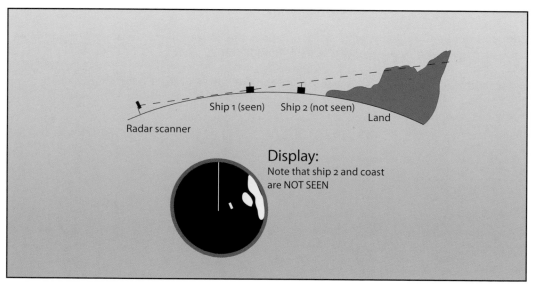

Figure 2.12   *The effect of radar horizon on the displayed picture*

What can radar see?

In order for the radar to see a return from a target, the object in question needs to be good at reflecting microwaves, and needs to reflect some of them in the right direction, i.e. back to the radar scanner.

As far as the target is concerned, a number of factors affect this:

■ **Size.** Obviously, the larger the object the more microwaves it can reflect back, so (all other things being equal) large objects are more detectable than small ones.

■ **Material.** Unfortunately for small boat users, two of the commonest construction materials – GRP and wood – are virtually transparent to microwaves. Metals (and other electrical conductors) are the best reflectors. Ice can be fairly poor, so high latitude sailors beware.

■ **Aspect.** Generally, a lot of the energy will be reflected in the way that light is reflected by a mirror. This is good news if it is reflected straight back to the radar, and bad news otherwise. Surfaces which are rough and uneven will scatter the waves, so will show up (albeit quite weakly) even if they are not at the best aspect to the radar.

Radar reflectors are made of metal, and they solve the aspect problem because, if you use a **corner reflector** shape, most of the energy is reflected off each of the different surfaces and travels back parallel to the direction it arrived from. That is why radar reflectors make better targets than their size would suggest.

Metal ships are generally good targets, because of the large number of metal surfaces (including the odd accidental corner reflector) and relatively large size.

As far as radar is concerned, the main factors are:

■ its power;

■ the distance over which it is trying to detect targets;

■ the **gain**, or amplification, of the reflected signal by the receiver.

The obvious tradeoff applies: a poor target will only be detectable at short range and high gain, whereas a good target will be detectable further away.

2

2

Examples of different types of radar target

Unlike range, it is very difficult to quantify how good various types of target will be – largely because of factors like aspect. So a cliff facing exactly towards you will be a much better target than the same cliff seen from the 'wrong' angle. The following table gives a very rough qualitative assessment of different types of radar target, assuming that all can be 'seen' above the radar horizon:

rocky shores / cliffs	good
built up areas	very good
large metal structures	very good
sloping sandy shores	poor
lighthouses	probably indistinguishable from surrounding land (*unless enhanced by RACON – see chapter 5*)
large metal ships	very good
small wood or GRP vessels	very poor
small vessels with radar reflectors	fairly good (but small)
buoys without radar reflectors	poor
buoys with radar reflectors	fairly good (but small)

Clutter

It seems rather hard that, after all this work and technology, we then complain that the radar sees too much for us! But of course it does. There are two types of return that we might rather like to get rid of, because they don't interest us. These are:

- returns from sea waves (**sea clutter**);

- returns from rain (**rain** or **precipitation clutter**).

The first of these is pretty well always a nuisance. We know there are waves around the boat and we don't need expensive technology to tell us that. On the other hand, returns from a rough sea can mask other returns, e.g. from small craft or navigation buoys, that we do want to see. Luckily, our radar can be asked to reduce the level of undesirable returns close to the boat (where the problem is greatest). Like the squelch control on a VHF radio, we don't turn the filter up too high or we will eliminate some of the targets we do want to see – more of this in Chapter 3.

Radar can do quite a good job of seeing heavy rain showers, and if we want to track where squall showers are, we sometimes use it for that purpose. But if we're more interested in seeing hard objects in the rain, particularly if the rain is all around us, we might prefer to reduce the rain clutter. Radar has controls to do this too, using slightly different techniques to those for sea clutter (and explained in Chapter 3).

How radar displays the picture

For the picture display, there are essentially two technologies. Old-style radar used an analogue display on a **cathode ray tube (CRT)** – similar to old-style television. Essentially the radar receiver was 'hardwired' to a display device which lit up the screen in the appropriate place. This arrangement gave limited flexibility as to how the picture could be manipulated and displayed. You can still see these displays in use, but for small craft they are being replaced by digital **liquid crystal displays (LCD)** and **raster** processing. This means that the picture is displayed in pixels from digital information stored and processed within the radar's processor, just like on a computer screen.

2

The advantages of the new display technology are significant. Apart from size and power (and cost) of the display hardware, from the designer's point of view you can do virtually anything with a picture which is stored and processed digitally within a computer – turn it round, zoom in on part of it, display data boxes, lines and symbols on top of it, write software to track targets on it, combine it with other displays, etc. As described above, you can use additional information to modify the picture: if your processor knows where the boat is pointing from one instant to the next, it can display a stabilized picture, either course-up or north-up, by simply rotating the picture information by the relevant number of degrees.

On the other hand, the cathode ray tube can still, in some ways, give a slightly better picture. For example, resolution and sharpness may be better than LCD displays (which have a limited number of pixels) and contrast may also be better. Also, some LCD displays give no indication of echo strength. This is a bit of a loss, particularly for seeing targets in clutter. Colour LCD displays are available which can address this limitation, and no doubt technology will continue to advance in this area.

Conclusion: looking at a radar picture

So now you know what you are actually looking at when you look at a radar picture. It's not magic and it's not perfect, but it can be very, very useful. If you turn the range up to, say, 24 miles, it just isn't going to show you every object in that range – some will be hidden behind other things, some will be below the horizon, and others won't reflect microwaves strongly enough, particularly at longer range. Concentrate on what *can* be seen, and work with that.

2

Points to note

- Radar sends out a narrow beam of microwave pulses, and detects echoes.

- The beam is wider in the vertical plane than in the horizontal plane.

- The picture shows range and direction of echoes that have been detected.

- Range is measured by timing the return of an echo.

- Direction is measured by knowing where the scanner (and therefore the beam) is pointing – the narrower the beam, the more accurate the direction.

- Stabilized pictures – course-up or north-up – get rid of the picture rotation when the boat yaws.

- Radar is 'line of sight', and has a horizon depending on antenna height.

- Radar cannot see targets which are hidden behind other objects.

- Things that are good at reflecting energy show up as good targets – wood and GRP do not, metal is good, large objects reflect more than small ones, and aspect matters.

- Unwanted returns – clutter – can be filtered out.

- Modern radars can process the picture in various ways, and displays are smaller, cheaper and consume less power.

3 Basic machine and controls – how to get a good picture

This chapter takes you through the basic controls that are available on any radar, and describes how you go about starting the thing up and getting a decent picture. Chapter 7 takes this a stage further by describing some of the more advanced functions available on modern radars, but you will be able to start using your radar with the controls described below.

Of course, the difficulty here is that, once you have got your own radar system, you will also need to study the *Owner's Handbook* to see how to operate its controls. In the old days, any radar would have had a row or two of knobs, clearly labelled with each of the functions I am about to describe. Not so now: in the interests of 'simplifying' the user interface and reducing the number of physical knobs and buttons, most new machines have a sophisticated interface using multi-function keys and software-driven menus. You will have no option but to familiarise yourself with this interface even if, in practice, you end up using only a small proportion of the machine's functions (an all-too-frequent situation in this modern world).

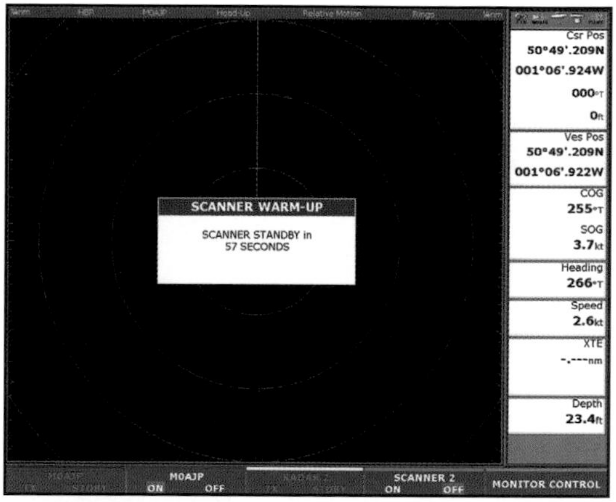

Turning the radar on

A **power** switch will initiate the start-up procedure. The radar function is not available

Figure 3.1 *Turning the radar on. Scanner warming up.*

Figure 3.2 *Adjusting the range scale.*

right away, because the transmitter magnetrons have to warm up which usually takes a minute or two. Generally, a message on the screen will tell you that this is happening. Note that with cathode ray tube sets, you may need to turn down the gain and brilliance to avoid damaging the screen (see the *Owner's Handbook*).

Remember, too, that the transmitter can be dangerous to anyone standing near the scanner: for the purposes of this book, it is assumed that you have read the warnings in the *Owner's Handbook*.

The radar will probably enter **standby** mode once the transmitter has warmed up. This means that it's ready to go but isn't transmitting and receiving yet. It's a good idea to keep your radar in standby when you are not actually using it, because it saves power while being available immediately when you need it.

You then select **transmit** when you want the radar to start operating.

3 Range scale control

This control is straightforward. A number of range scale settings will be available which determine the scale of the display, by reference to the maximum range displayed, i.e. the distance from the centre of the display to the edge. Typically, the smallest range scale setting will be a fraction of a mile, and the largest 16 or 24 miles, and on some sets you can even customise your own range scale settings.

Range rings (concentric rings at set intervals on the display) are helpful, but they can be turned off if you want a clearer picture. Leave them on while you are adjusting the display. For each range scale setting, the display should tell you the range to the outer edge of the picture, and the range ring interval.

Brilliance, or light and contrast

You can adjust the display so that things can be seen as clearly as possible in the prevailing light conditions. With LCD displays, you will generally be able to adjust the backlighting and the contrast. This is similar to other devices' LCD displays: GPS, for example. Normally, LCD displays come up with reasonable default settings, but given that you want to get the most you can out of the radar picture, it is worth spending a little time adjusting for best clarity and comfort.

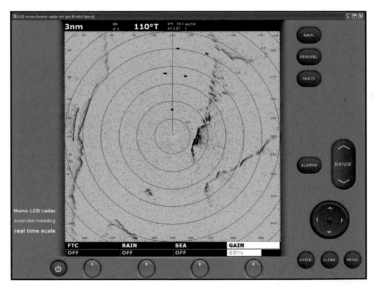

Figure 3.3 *Use of 'soft' controls. Selection is by the 'soft key' buttons beneath the display, and adjustment is by the 'track pad' to the right of the display. Here, Gain is adjusted to 65%.*

With cathode ray tube displays this adjustment is a bit more important, because it can affect whether small or indistinct targets can be seen on the screen. You need to adjust brilliance – literally how brightly the display is painted. Look at the range rings and adjust them so that you can see them clearly and distinctly: too bright and they become blurry and dazzling, too dim and you can't see them clearly. On some sets, there may also be a focus control which you adjust to get the lines nice and sharp.

Gain

Gain is like a volume control for the radar receiver.

As described in Chapter 2, the radar *transmits* microwave pulses (at quite a high power level) and then *receives* the very faint echoes of those pulses so that it can display them for the user. The receiver is rather like a radio receiver. It is tuned into a particular frequency (see next section on tuning), and it picks up small signals and amplifies them. With the gain turned right up, all the **noise** that the receiver detects will be treated as returns; but when turned right down you will get hardly any picture at all, because all but the strongest returns will be missed.

Modern sets have an automatic option for gain, and this will probably do a better job than you can to begin with, if you are inexperienced. A few default settings are usually provided depending on where you are operating the radar: in harbour, open sea, etc.

But it is important that you understand this control and have a go at adjusting it yourself. It is fairly critical to have it correctly adjusted or you will miss small targets. Also, you may want to 'tweak' this adjustment up and down when you are using the radar and trying to find targets.

3

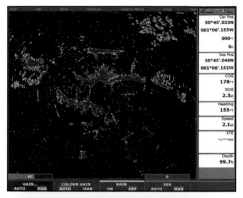

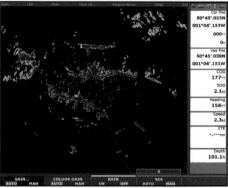

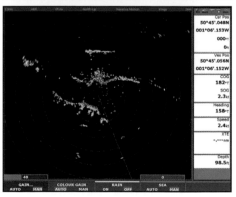

Figure 3.4 *Three screenshots: very high gain (top); auto setting (middle); very low gain (bottom). Note that some targets are missed with low gain setting, and high gain setting includes a lot of noise.*

The correct setting is to have a slight 'speckle' (i.e. very slight amount of noise) all over the screen so that you are not missing any targets. You need to check this setting when you change range scale, and you may need to reduce the level of noise to make small targets more distinct.

Tuning

This control tunes the receiver to the exact frequency of the transmitter. Obviously the receiver will be roughly tuned in already because the manufacturer knows what frequency the transmitter is specified to transmit on, but fine tuning is important because the transmitter frequency can drift slightly.

The tuning control will probably also have an automatic option, and in this case it is probably the better one – not least because the receiver may need retuning as everything warms up. If manual tuning is necessary, the display will probably give you a special indication of signal strength in the form of a segmented bar, and you tune for the strongest signal. In the absence of this, you tune by looking at a weak return and getting it as bright as possible.

The display

At this stage you have a radar picture. The above controls are all you need to know about in order to achieve this.

The basic radar picture – known also as the **plan position indicator**, or **PPI** – shows your own ship's position at the centre; and running vertically up the display from the centre is a straight line called the **ship's heading marker**. In general, the default picture will also have a series of equally spaced circles centred on the ship's position: these are called **range rings**. (Note that both the range rings and the ship's heading marker can generally be turned off.)

Radar returns are displayed on the picture at the *angle* (from the ship's heading) and the *range* (from the ship) at which they are detected. Thus the picture is a plan view of the radar returns detected around your vessel, with the ones dead ahead shown at '12 o'clock' on the display. This basic display orientation is called **ship's heading-up**.

The remaining sections in this chapter describe:

- improvements to the picture; and

- simple facilities that make it easier to measure the position of targets and keep track of them.

Sea clutter

As explained in Chapter 2, you tend to get returns from the water around you, particularly off waves in rough conditions. These are perfectly valid as far as the radar is concerned: microwaves reflect reasonably well off water, and the radar is detecting lumps of it – particularly close to the boat, where the beam is angled downwards (it is quite wide in the vertical plane, remember) and tends to get better reflections from the surface of the sea. However, the detection of such targets is not useful to the observer, who knows that he is surrounded by water but wishes to detect small hard objects floating in it.

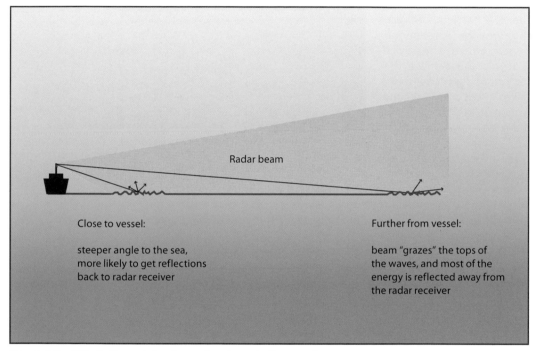

Radar beam

Close to vessel:

steeper angle to the sea,
more likely to get reflections
back to radar receiver

Further from vessel:

beam "grazes" the tops of
the waves, and most of the
energy is reflected away from
the radar receiver

Figure 3.5   *Effect of sea clutter is greater closer to the vessel*

Radar designers tackle this problem by giving you a facility to reduce the gain at *short range only*. Further out, where the clutter isn't such a problem, the gain is restored so that you will see weak targets. The sea clutter control can be adjusted by the user, or an automatic option is provided on some sets.

As with gain, by all means use the automatic option, which again may have several default settings to choose from. But the advice is to have a go at controlling it yourself, so that you can learn how to operate it and see what it does for you. Use the manual control to adjust up and down a bit when you are trying to spot small targets in the clutter. Once you've seen what the manual control can do, use automatic if you are satisfied with what it does.

Using this control is a bit of an art. Turned up too high and you will lose targets close to your boat because they will be suppressed along with the sea returns. Too low and you won't see the real targets because they will be surrounded by sea returns. The advice is to try and achieve the background 'speckle' that I talked about above (see the section on gain), so that you don't lose any targets that are as strong as, or stronger than, the sea returns (if they are weaker, you will never detect them anyway).

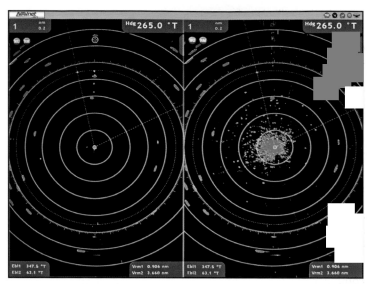

Figure 3.6 *Effect of sea clutter control: clutter is suppressed on left picture, unsuppressed on right. Left picture is clearer, but have any real targets been eliminated?*

Remember, what you are looking for is a return that is consistent over several scans – returns from the sea will be random and inconsistent from one scan to the next.

Rain (precipitation) clutter

Rain (or snow or hail) is a problem for radars, because the drops or particles do reflect and scatter microwaves (as they do with light). So the precipitation will show up as unwanted returns, potentially masking 'real' targets. This is called **rain** or **precipitation** clutter. Unlike sea clutter it can occur anywhere, not just at close range.

There are techniques to suppress this type of clutter, one of which exploits the characteristic of this sort of return: that it is continuous over a spread of ranges (whereas a typical target return is much shorter). If the radar can process detected returns so that only the leading edge is displayed, it can improve the visibility of targets in rain (and in some circumstances improve the discrimination between two targets close together in range that might otherwise merge together). This control is also referred to as **fast time constant** or **FTC**.

The same principle (as for gain and sea clutter) applies to using rain clutter controls: the user needs to adjust the settings carefully to improve clarity but avoid suppressing real target returns. FTC will also affect the way other targets (e.g. land) are displayed, so may be better left off when not needed for rain.

Precipitation will also absorb or scatter energy from microwaves that are detecting targets *beyond* the rain, making it harder to detect such targets. This is not actually a *clutter* problem, and the solution is to increase the gain, or to use any other facility for enhancing or stretching targets (see Chapter 7).

Electronic bearing line

The above sections are all about improving the radar picture towards achieving an ideal situation where all the 'real' targets, strong or weak, can be clearly seen. Having done that, you will want to exploit the radar to track targets and measure where they are. Two controls in particular are provided for this purpose (and in fact they are sometimes combined in a single cursor).

The **electronic bearing line (EBL)** is a line that can be displayed on the screen, from the centre outwards, indicating a particular bearing from your boat (which you can adjust). For example, if you place this line over a target and the target stays on the line as it comes closer, then the steady bearing of that target indicates that you are on a collision course. Also you can read off the actual value of this bearing (relative to the boat's heading or, if you are using north-up display, the true bearing).

It is useful to familiarise yourself with this control on your radar.

Figure 3.8 *Electronic Bearing Line (EBL) and variable range marker (VRM). See magenta and red lines on the display. Two EBLs and VRMs are displayed.*

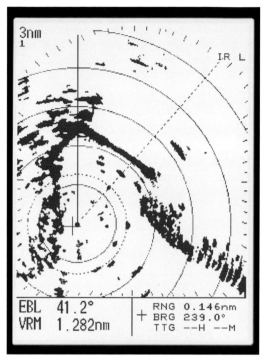

Figure 3.7 *Electronic Bearing Line (EBL) and variable range marker (VRM). See dotted lines on display.*

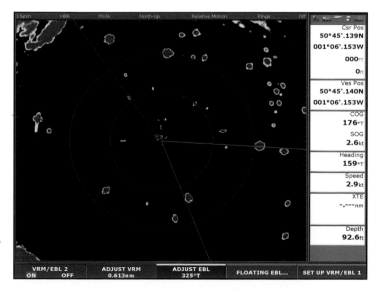

Variable range marker

In similar vein to the EBL, the **variable range marker (VRM)** is an adjustable range ring which you can use to measure the range of a target and/or track the change in range. Again, it is useful to be familiar with this control's operation.

Cursor

On typical modern radars the cursor will give you a read-out of range *and* bearing in a single action, as well as enabling you to set the VRM and EBL.

Practical exercises

1 Getting a useable picture

On a calm day with good visibility, when you have plenty of time, take your boat out of the marina and into open water away from other traffic.

Let the boat drift (safely!) while you turn on the radar. The set will warm up and enter standby mode.

Select a range scale of, say, 3 miles – so that you will have targets (other boats, land, the marina) within that range. Make sure the range rings are turned on (this is normally the default).

Start the radar transmitting.

Adjust your display so that it as clear and visible as possible, using the display light and contrast controls (assuming you have an LCD display).

Find the gain control, and select auto (probably the default). Find tuning and select auto (probably the default).

At this stage you should have a radar picture. Take some time to reconcile your radar picture with what you can see with your eyes. Check out major features and returns to begin with – then have a look for small targets (other small boats).

Go back to the gain control and select manual. See what happens when you adjust it up and down. Try to find the optimum level (slight speckle in target-free areas of the picture). Compare this with the picture when you select auto – are any small targets missing?

Go to the sea clutter control (which will probably be set to auto too), select manual and reduce it to zero. A bit of sea clutter may appear (unless it is very calm). Increase the clutter suppression and see the clutter reduce. Try overdoing it, and see if genuine targets fade or disappear. Select the optimum level (very slight clutter) and compare this to the auto setting.

Repeat any and all of the above as often as you like for practice – try positioning the boat fairly close to a known target (buoy or other small craft), and try other range scale settings.

Remember – this is the very start of a long learning process, so be patient! Things may be confusing, and you may not be able to figure everything out on this first sea trial.

2 Working with a target

On a good, clear day repeat the above steps to the point where you have a satisfactory radar picture.

Select another small vessel. Observe where it is and what it is doing, and identify it on the radar (ideally get a friend to act as the target, and stay in communication with them).

Mark the target with an EBL, and work out the bearing (you can check it roughly with a compass).

Mark the target with a VRM and note its range.

Watch the target move against the original marked position.

With a cooperating boat, approach on a collision course, and note the target moving down the EBL.

The aim of this is to get used to identifying targets and tracking them with the EBL and VRM – and getting used to your radar's EBL/VRM controls. Repeat until you are happy, and do this regularly in GOOD VISIBILITY before you are caught out in fog!

3

4 How to use radar for collision avoidance

Introduction

Collision warning and collision avoidance are really major applications for radar – in fact they are where radar out-performs any alternative technology. In poor visibility, radar is literally the only device that can detect other vessels (and large floating objects); it will tell you how far away they are and in what direction, and enable you to track where they are going. Your chart plotter and GPS will be invaluable for navigation, but they just don't know about other vessels.

In good visibility your main tools to assess risk of collision are:

■   keeping a good visual look-out;

■   keeping track of other vessels' bearings.

With binoculars, you have the advantage of being able to observe what other vessels are doing. The drawback is that you cannot measure their range accurately. Radar can supply this information, and it can also assist the look-out as an alternative means of spotting other vessels.

In poor visibility, or at the limits of visibility, the visual look-out is severely disadvantaged, and radar is your main means of spotting and tracking other vessels.

Of course, with or without radar, low visibility is a serious hazard for all shipping. Do not think that, because you are equipped with radar, you can safely go out in fog. Unless you are a very well trained and practised radar observer, it is all too easy to:

■   miss small targets (even if they have radar reflectors); and

■   become very confused while attempting to interpret the actions of – and avoid – larger ones.

It is basic good seamanship to exercise extreme caution with fog, because it is a serious hazard.

View radar as a useful *emergency* tool in low visibility, and by all means use it in *reduced* visibility, together with a good and effective visual look-out. Radar can assist your visual look-out, because it will enable you to spot more distant targets and give you accurate distance and movement information. On the other hand, visual observation will give you information (particularly shapes and lights) that is not available from the radar display. Both types of observation will assist in decision-making for collision avoidance, and using the radar in this way will enable you to develop your radar skills for any occasion when you are caught out in fog.

AIS

There is one other system that is able to give you position information for *some* other vessels, and it is becoming an increasingly popular and valuable tool for leisure craft skippers. **Automatic identification system (AIS)** provides a very useful service, which is complementary to radar and visual look-out. AIS transponders are mandatory for large vessels and are becoming more common on smaller ones. They send messages with information about the ship's position, identity and movement on a dedicated radio channel in the VHF band. These messages can be picked up by other vessels equipped with receiving equipment (even if they are not fitted with transponders). As well as displaying the full message, the receiving equipment can present information graphically on radar and chart plotter displays.

From a technological point of view, AIS has nothing to do with radar. But because of the ability of modern instruments to 'talk' to each other, it can be usefully integrated with radar to identify targets and aid collision avoidance. Further details are given at the end of this chapter.

4

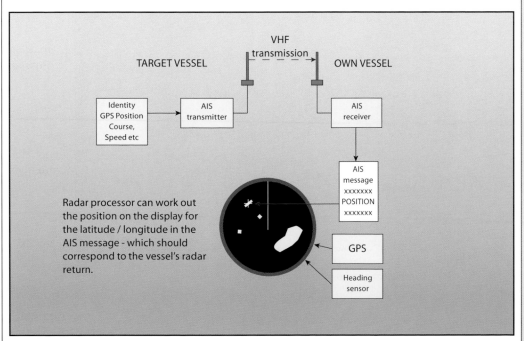

Figure 4.1 *How AIS works. Position information in the message from the transmitting vessel can be matched to a particular radar target by the receiving vessel.*

Practical approach in reduced visibility

First turn on your radar! Go through the process described in Chapter 3, paying particular attention to the adjustments (gain, clutter) that will help you to see small targets. Go back to those adjustments from time to time while you are using the radar – adjusting them up and down can help to reveal targets.

Remember the **International Regulations for Preventing Collisions at Sea (IRPCS)**, Rule 5, Look-out: 'Every vessel shall at all times maintain a proper look-out by sight and hearing as well as by *all available means* appropriate in the prevailing circumstances and conditions…'. This, of course, includes radar.

Incidentally, make sure that your own radar reflector is deployed. (Some people keep them in the locker, for reasons I have never understood.) If using radar achieves no other purpose, it should make you sensitive to the difficulties that other people probably have when trying to observe *your* vessel.

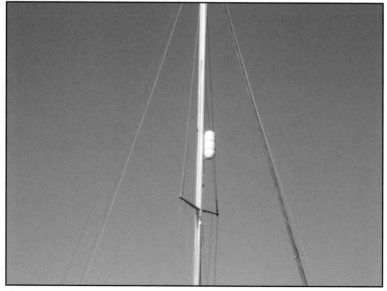

Figure 4.2 *Sensible place for a radar reflector. The one shown is a metal corner reflector type in a plastic housing.*

About radar reflectors

There is endless discussion about the effectiveness of radar reflectors, largely because radar's ability to detect small craft is affected by so many unpredictable factors that the whole question becomes very confused and is beset by contradictory anecdotal evidence and claims.

In the wake of the loss of the yacht Ouzo, the UK's Marine Accident Investigation Bureau (MAIB) commissioned a study into the effectiveness of various devices that are commercially available and suitable for small craft. The report makes very useful reading (see References).

The research tests a number of devices in controlled conditions, but it is worth remembering that they are generally *used* in very *uncontrolled* conditions. Factors such as how the devices are installed, and the heel angle and motion of the vessel, can have a huge effect on how well radar reflectors perform. Add to this the factors affecting the observing ship (skill and attention of the observer, sea state and clutter, range, clutter suppression and other adjustments, use of S-band or X-band radar) and you can see that the situation is far from simple or predictable.

Devices can be divided into **active** and **passive**.

Passive devices are simply reflectors of the transmitted radar signal. There are two families of passive reflectors:

■ those that use metal **corner reflectors**, i.e. three metal planes arranged at right angles to each other. This arrangement means that microwaves will be reflected back parallel to the direction of the incoming radar beam, i.e. back towards the scanning radar;

■ those that use the **Lunenburg lens** principle, a radar lens which focuses the incoming microwaves, reflects them, and directs them out through the same lens parallel to the direction of the incoming radar beam.

The important thing to know about both of these is that:

■ they are both affected by the angle (azimuth and elevation) from which they are scanned, and therefore by the attitude (roll, heel and pitch) of the vessel they are attached to. They need to be installed correctly to alleviate this. It is possible for a good Lunenburg lens device to be uniformly effective through 360° of *azimuth*: not possible with corner reflecting devices;

■ size matters: what you are trying to simulate is a radar target consisting of a large area of metal, so it helps if you employ the largest possible reflector to achieve this.

Active devices (or radar transponders) *detect* the radar signal, and *transmit* a return which will generally be stronger than any reflected signal. Of course active devices need power to do this, and may be effective on a limited range of frequencies (i.e. X-band or S-band).

4

4

They tend to give better target enhancement performance, but may also be affected by the angle (particularly elevation) of the scanning radar.

Another defence is to carry an AIS transponder (this is a radio transmitter device, *not* a radar transponder) – see the section on AIS at the end of this chapter.

Finally, there are devices that simply warn you, by sounding an alarm, when the vessel is being scanned by radar (some active devices do this too), and tell you from which direction you are being scanned. This is useful in an area where shipping is sparse, particularly if you are sailing short-handed. It is not much good if you are constantly being scanned by several radars.

Select the appropriate range for detecting targets.

You can use your own experience to decide the range at which you want to become aware of other vessels, depending on your circumstances. In coastal waters, for example, you are more likely to be looking for small vessels, which may only show up as radar targets relatively close to you, and you are not going to start manoeuvring to avoid small vessels 5 miles away. So the appropriate range scale is probably around 2 miles. On the other hand, in open water (for example, crossing the English Channel), you might want to start looking for larger ships at a range of 8–10 miles. Because they are large, they will be good radar targets and are likely to show up at that distance. They will probably be following a steady course, and observing them for longer will enable you to determine better what they are doing. And they may be fast moving, in which case you need to be aware of them early.

In some circumstances, such as in really poor visibility, you may be looking out for both small and large targets, so you will need to switch range, or use the dual range facility available on some radar systems. This requires some skill and is time consuming, so it may be preferable (if conditions allow) to use visual lookout for small targets close to your vessel, and use the radar for larger, more distant ones.

Figure 4.3 *Fast target – thankfully, in good visibility.*

Fast targets

It is worth scaring yourself a bit by working out how quickly a high-speed ferry can close on your vessel from the time you first detect it. If your set is on a 2-mile range scale and your own speed is 7 knots, a vessel travelling at 45 knots on a reciprocal course will appear on the display for 2 minutes 18 seconds before it reaches you. If your own speed is 15 knots, the corresponding time is exactly 2 minutes. This underlines the necessity of keeping a vigilant radar watch (think how easy it is, as skipper, to be distracted for a minute or two), and of spotting large targets at greater range if you are to have a chance of working out what they are doing, and how – or whether – to manoeuvre to avoid them.

4

On passage, try to hold a steady course and speed. This will make the radar picture easier to interpret (and your own actions easier to interpret by other vessels).

Collision risk

The key to determining whether a **collision risk** exists is to see if the bearing of the observed vessel changes as it approaches. In good visibility this is determined by taking a compass bearing on the vessel every few minutes. With radar, you monitor the vessel's bearing with an electronic bearing line (EBL). This works with a basic ship's heading-up, or azimuth stabilized course-up or north-up display. If you are using the basic ship's heading-up display, it is important to keep your vessel on a constant heading during this process. (Note: this works with **relative motion** displays, where your own ship is at the centre of the picture, but does *not* work with **true motion** displays. See Chapter 7.)

Once you have a target, place an EBL on it and monitor to see if it comes down that bearing line or starts to go off it. At the same time, tell the look-out to try and spot the vessel on the relevant bearing. If your radar has the facility, you will probably be able to monitor several targets in this way using several EBLs.

At this stage, with many of the targets you spot, you will probably be able to reassure yourself that they present no threat, because they change bearing (albeit slowly at long range). Just keep your own vessel on a steady course and watch carefully – don't start to make tentative course changes 'to improve the situation' when the other vessel is still several miles away. It is all too easy to make a wrong assumption about what the other vessel is doing and make the situation worse, not better. Continue to monitor 'safe' targets until they are well clear, in case one of them alters course or speed and suddenly becomes a more immediate danger!

4

Radar assisted collisions

Despite the widespread use of radar, vessels continue to collide, and there have been tragic instances where two vessels equipped with radar have managed to engineer a collision which would never have happened without it. A typical scenario is as follows:

- Ship A, heading due north, observes a target fine on the starboard bow, at relatively long range. Because of the range, the bearing changes very slowly, so the skipper decides that the other ship is heading south and will pass down his starboard side, but rather too close. He alters course 5 degrees to port, and instantly the radar picture becomes much more encouraging, because the target is a lot further away from the ship's heading line.

- Ship B, the target vessel, is in fact heading slightly west of south, and therefore observes Ship A fine on his port bow. For similar reasons to Ship A, his instinct is to 'turn away' from the target and he alters course to starboard.

- Gradually, both skippers notice the target vessel getting closer to their heading line again, so they each repeat their previous course alteration.

- This process continues until they collide!

Having personally made a similar mistake, but fortunately with a happier ending (I was unable to catch up with the ship to collide with it), I am very aware of how confusing this kind of situation can be. Your own course alterations make it far harder to work out what is going on. The answer is to observe the other vessel for longer without making any course alterations, and not make assumptions about what it is doing, particularly at long range when you simply don't have enough information.

Now let's assume that you have a target that continues to track steadily down the EBL. It's a given that you will ask the look-out to keep a particularly sharp eye in the target's direction, because the sooner you can see what it is and what it is doing, the better. But if you still can't see it, you can use the radar observations over a period of time to determine:

- the target's course and speed; and

- the **closest point of approach (CPA)** if both vessels continue on their present course and speed.

International Regulations for Preventing Collisions at Sea

At this point we should consider what the Collision Regulations say about restricted visibility. A number of the rules apply to vessels 'in sight of one another'. This means they do not apply when the ships can only observe each other by radar. The rules that come under this heading are:

- Rule 12 – Sailing vessels

- Rule 13 – Overtaking

- Rule 14 – Head-on situation

- Rule 15 – Crossing situation

- Rule 16 – Action by give-way vessel
(but note that Rule 8 – Action to avoid collision still applies)

- Rule 17 – Action by stand-on vessel

- Rule 18 – Responsibilities between vessels

Instead, there is a separate Rule 19 – Conduct of vessels in restricted visibility.

4

IRPCS Rule 19

Rule 19 – Conduct of vessels in restricted visibility

a) This Rule applies to vessels not in sight of one another when navigating in or near an area of restricted visibility.

b) Every vessel shall proceed at a safe speed adapted to the prevailing circumstances and conditions of restricted visibility. A power-driven vessel shall have her engines ready for immediate manoeuvre.

c) Every vessel shall have due regard to the prevailing circumstances and conditions of restricted visibility when complying with the Rules of section I of this part *[i.e. Rule 5 – Look-out; Rule 6 – Safe speed; Rule 7 – Risk of collision; Rule 8 – Action to avoid collision; Rule 9 – Narrow channels; Rule 10 – Traffic separation schemes].*

d) A vessel which detects by radar alone the presence of another vessel shall determine if a close-quarters situation is developing and/or risk of collision exists. If so, she shall take avoiding action in ample time, provided that when such action consists of an alteration of course, so far as possible the following shall be avoided:

i) an alteration of course to port for a vessel forward of the beam, other than for a vessel being overtaken;

ii) an alteration of course towards a vessel abeam or abaft the beam.

e) Except where it has been determined that a risk of collision does not exist, every vessel which hears apparently forward of her beam the fog signal of another vessel, or which cannot avoid a close-quarters situation with another vessel forward of her beam, shall reduce her speed to the minimum at which she can be kept on her course. She shall if necessary take all her way off and in any event navigate with extreme caution until danger of collision is over.

This is a bit of a change from what we're used to. What it amounts to is a shift of responsibility; *all* vessels are required to take action to avoid collisions on a more or less equal basis. Gone are the definitions of which vessel is to give-way and which to stand-on, in the various different situations covered by Rules 12 to 18. To a large extent this is very logical, because a vessel can't observe what the other vessel is, or what it is doing; but it is certainly a big change to the normal rules and responsibilities.

The first thing to note about Rule 19 is our obligation, when observing another vessel by radar alone, to 'determine if a close-quarters situation is developing and/or risk of collision exists'. In case we have any doubt about how we should go about this, it is worth studying Rule 7 – Risk of collision, and noting in particular (b) and (c): you need to make 'proper use of radar equipment', undertake 'systematic observation of detected objects' and *not make assumptions 'on the basis of scanty information, especially scanty radar information'*. You have been warned!

IRPCS Rule 7

Rule 7 — Risk of collision

a) Every vessel shall use all available means appropriate to the prevailing circumstances and conditions to determine if risk of collision exists. If there is any doubt such risk shall be deemed to exist.

b) Proper use shall be made of radar equipment if fitted and operational, including long-range scanning to obtain early warning of risk of collision and radar plotting or equivalent systematic observation of detected objects.

c) Assumptions shall not be made on the basis of scanty information, especially scanty radar information.

d) In determining if risk of collision exists the following considerations shall be among those taken into account:

i) such risk shall be deemed to exist if the compass bearing of an approaching vessel does not appreciably change;

ii) such risk may sometimes exist even when an appreciable bearing change is evident, particularly when approaching a very large vessel or a tow or when approaching a vessel at close range.

On this basis, we certainly need to take seriously the plotting of radar targets.

Plotting

We can plot the position of a target, relative to our own vessel, over a period of time. From this plot, knowing our own speed, we can work out the course and speed of the target vessel, and its aspect to us: are we seeing its port/starboard bow, beam, quarter or stern? And assuming that both vessels hold their current course and speed, we can also work out what the **closest point of approach (CPA)** with the other vessel will be, and when that will occur (the **time to closest point of approach – TCPA**).

The target tracking and plotting process can be automated. Equipment to do this (**automatic radar plotting aid – ARPA**) has been available on large ships for some time, and these functions are now becoming available on small craft radar systems. This is undoubtedly a huge benefit, and anyone who has such a function on their radar system will naturally prefer to use that rather than laboriously plotting targets. But there is considerable merit in understanding what the automated process is doing, because you do need to be aware of its limitations. Essentially, what it's doing inside the processor is exactly what you do when plotting on paper, so it's worth reading this section to understand the process (even if you never intend to do it the old fashioned way).

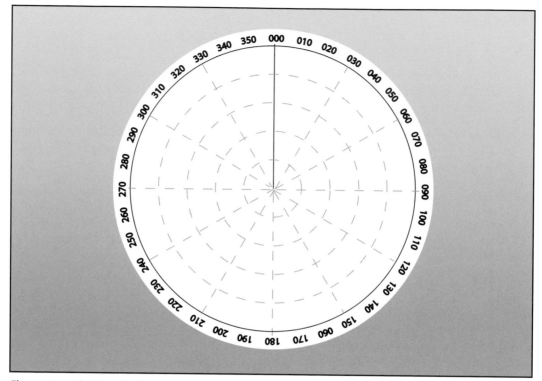

Figure 4.4 *Plotting sheet*

If you are doing it by hand, you can mark the screen with wax (chinagraph) pencil or use a paper plotting sheet (you can buy these, or make your own along the lines of the blank one included above). The latter has the advantages of being more accurate and allowing you to change the range scale – and keeping the screen clean! Even professional radar operators need to practise paper plotting (in case their ARPA goes wrong), so we might as well follow the same procedure as they do. If you are inexperienced, it certainly helps if two people can work on the radar plotting together – one observing the radar and reading off distances and bearings, the other plotting – while a third person helms the boat and keeps a look-out.

Plotting method

The following method of plotting can be done from a basic display (ship's heading-up), or azimuth-stabilized course-up or north-up display, but *not* from a true motion (ground or sea stabilized) display. (See Chapter 7 for a detailed explanation of stabilized displays.) To avoid confusion, it is recommended that you construct the plot in the same orientation as the display. Ship's heading-up or course-up probably gives you a better feel for the situation as it develops ahead of your vessel and, of the two, the stabilized course-up, if available, will give you a steadier picture. Use a range scale setting which gives you a clear view of the targets you are plotting: shorter range settings give better measurement of range and bearing.

1. On the plotting sheet, draw in your own ship's heading line. Assuming you are using ship's heading-up or course-up, this is a vertical line straight up the sheet from the centre. *[For north-up: own ship's heading line will be in the true direction of the heading, not vertically up the display.]*

2. Mark the position of the target on the sheet at 3 or 6 minute intervals. Label each plot with the time in minutes. (Why 6 minutes? Because it is one tenth of an hour, and that keeps the maths easy. A vessel doing 7 knots moves 0.7 miles in this time.)

3. When you have a few positions in a reasonably straight line, draw a straight line through these points and extend it until it passes the centre of the screen. This is the path the target will follow if you both continue on the same course and speed.

4. Measure the closest distance between this line and the centre. This is the target's Closest Point of Approach (CPA).

5. By stepping out the target's progress down this line with dividers, you can gauge the time at which this closest approach will occur (TCPA).

6. From the target's first position, draw a line parallel to your own ship's heading line, going vertically down the display. This is the line that the target would have followed *if it were stationary*, so that its only movement on the display would have been due to your own vessel's movement. *[For north-up: line will be in the reciprocal direction to the heading line drawn in step 1, not vertically down the display.]*

7. As you plot the target 12 minutes after the first position, calculate how far your vessel has travelled in that time, e.g. 2 miles if you are doing 10 knots.

8. Measure off this distance down the line that you drew in step 6, from the target's first position. Mark this point W.

9. From W, draw a straight line to the target's 12-minute position. This line shows the direction the target vessel is travelling in relative to your own heading, and the distance from W is the distance it has travelled along its track in 12 minutes (multiply by 5 and you have its speed). From the angle of its track, you can also judge what aspect the vessel is presenting to our own. *[For north-up: this line will show true direction of the target vessel's course.]*

4

Examples

Figure 4.5 shows the process described above. In this example, the CPA is half a mile and will occur about 18 minutes after the last plot. The target will pass just behind us. Our speed is 10 knots (hence W is 2 miles down the magenta line). The target heading is 30 degrees to starboard in relation to our own heading, and he is moving at just over 6 knots. We are just aft of his starboard beam, and appear to be overtaking him.

4

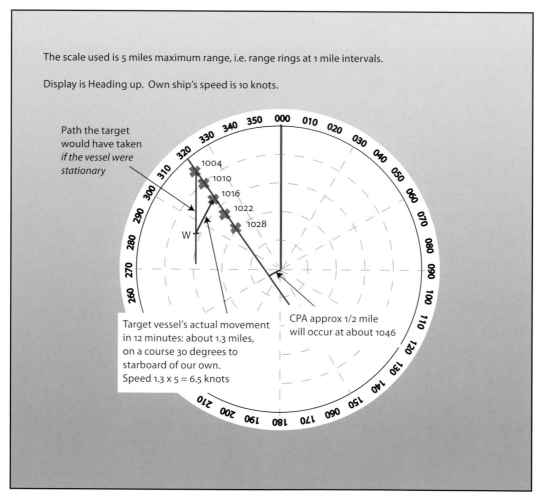

Figure 4.5   *A single target, which will pass just behind us.*

The second example shows a situation with three target vessels. Our speed is 8 knots. It is instructive to look at the figure first *without* performing the plotting process (see Figure 4.6): how easy is it to estimate what each of the other vessels is doing? Can you even guess whether each vessel is moving faster or slower than our own?

Figure 4.7 shows the answer. Each vessel is doing something very different, with the yellow target crossing nearly at right angles to our own course, the blue on a near reciprocal course and the green overtaking on a near parallel course to our own. Yellow is doing 17 knots, blue 5 knots and

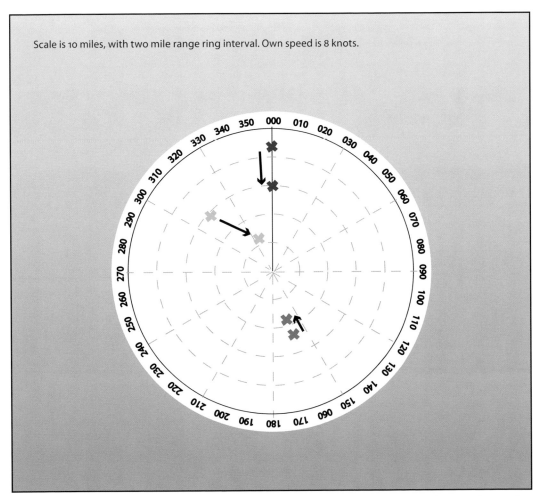

Scale is 10 miles, with two mile range ring interval. Own speed is 8 knots.

Figure 4.6 *Situation with three targets, showing initial positions and new positions after 12 minutes. Can you guess what each vessel is doing? Which vessels are moving faster/slower than our own?*

green 13 knots. You probably guessed that the green overtaking vessel was going faster than our own, but how about the other two?

We can deduce little or nothing about the actions of another vessel from a casual observation of the target on the screen. After performing a plot, manually or automatically, we are a great deal better equipped to decide what to do, under Rule 19 (Conduct of vessels in restricted visibility) and Rule 8 (Action to avoid collision). Note that in deciding what manoeuvre to make, we need to consider other targets and any restrictions to our own navigation.

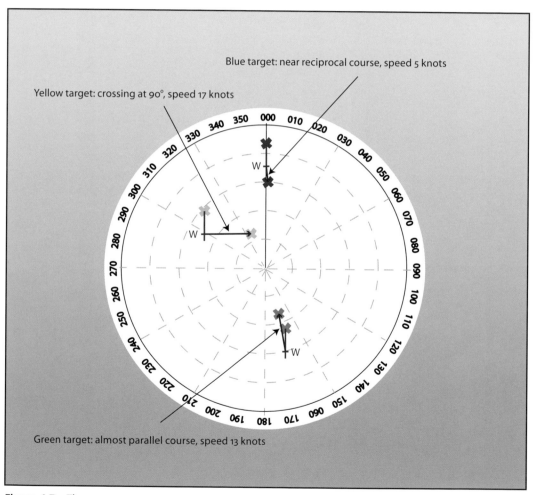

Figure 4.7   *The answer*

Also we need to continue to monitor this target after our manoeuvre to make sure it does what we expect it to do. In the example in Figure 4.5 we might decide to run more parallel to the other vessel for a while as we overtake, in which case it should eventually pass further behind us. But if the other vessel also changes course and/or speed (e.g. alters course to starboard) we might still get into a close-quarters situation, despite our own course alteration.

Automated target tracking

The automation of this process is described in a bit more detail in Chapter 7. Sufficient to say here is that it can:

- track a number of targets automatically;

- for each one, give the CPA and TCPA (assuming neither vessel alters its course and speed);

- give the target vessel's course and speed; and

- give a graphical indication of the target's course, speed and track on the display.

VHF radio

If you find yourself manoeuvring to avoid a close-quarters situation with another vessel, and are in any doubt, it is sensible to call them on the VHF radio. 'Any doubt' definitely includes a situation where you can't see each other!

However, another word of caution is required here. The danger is that you may not be talking to the vessel you think you are. When you call, anyone can answer, and a dangerous misunderstanding could arise if you are talking to the wrong vessel, particularly if you cannot see each other and confirm what your vessel and theirs look like. Clearly this difficulty depends on context (e.g. how many vessels are in the area), and AIS (if fitted on the other vessel) could definitely help you to identify who you wish to talk to.

If your radar is integrated with the GPS, you should be able to put the cursor on the target and get a read-out of its position (latitude and longitude), which helps when you are calling them ('Ship in position... this is sailing yacht Aztec 2 miles on your port bow', etc).

Commercial vessels seem to be quite happy with this type of contact, because it makes their task easier and you can agree what each vessel will do.

Automatic Identification System (AIS)

As mentioned at the beginning of this chapter, AIS is not radar technology, but it has been recently introduced mainly as a requirement for large commercial vessels, to give a receiving vessel (or shore station) real-time information about other vessels in the area. This information can be used to aid collision avoidance and can be of great assistance to small craft.

Messages are transmitted automatically on a dedicated VHF radio frequency, and are received by dedicated receivers – and as a small leisure craft you can fit a receiver on your vessel even if you are not required to carry a transponder. This is becoming increasingly popular. Transponders suitable for small craft are also available, and could help to protect you by making your vessel more noticeable to large ships.

If you receive another vessel's AIS message, it will include (among other things) the following information:

- vessel's identity – MMSI number for direct calling by VHF DSC, name and call-sign;

- size and type of ship;

- position;

- course;

- speed;

- heading

This message will be repeated every 12 seconds or better for a ship not at anchor (3 minutes if it is at anchor). Your on-board integrated instruments can use the ship's position to match this message to a particular radar target. This enables you to:

- confirm the target vessel's course and speed, deduced from your plotting exercise or automatic tracking; and

- call the vessel direct using its name, call-sign and/or MMSI number.

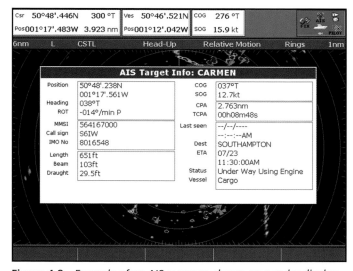

Figure 4.8 *Example of an AIS message shown on a radar display.*

While it may not remove the need for systematic observation of targets in restricted visibility (or, indeed, make navigation in low visibility very much safer – there are plenty of fishing vessels, leisure boats and small commercial ships which are *not* transmitting, remember), this facility will certainly help you to monitor the target vessel more quickly and accurately, particularly if it changes course – you will get much earlier notification of this than you could by plotting, automated or otherwise.

AIS is particularly useful where the vessels you are mainly concerned with are big ships transmitting AIS messages: going across the English Channel, for example. In these circumstances it may remove the need to plot some or all of the observed targets.

Practical exercises

1 Spotting small targets

In good visibility, take *every* opportunity to turn on your radar and adjust it to detect small targets.

Systematically compare what you can see visually with what you can see on the radar.

Find the range setting(s) that you are comfortable with for your normal speed and area.

If you can see a target that isn't detected by radar, try the gain and clutter adjustments at various range settings and/or make a mental note of the type of target that just *cannot* be detected (does it have a radar reflector?).

2 Assisting the look-out in moderate visibility

In any conditions where visibility is less than ideal, and distant vessels cannot be seen clearly, use the radar as a back-up to the visual look-out – particularly to warn the look-out of targets that they cannot see yet (tell them where to look).

Assess the actual visibility by noting the range of objects when they can first be seen.

At night, use the radar to check the range of visually-observed targets – this helps, because it is very difficult to judge distance from navigation lights alone.

3 Blind navigation exercise

Personally, I'm not a great fan of the rather artificial blind navigation exercise much beloved by some Yachtmaster® Examiners. However a *realistic* exercise, making full use of the available instruments, is a different matter – because you are practising what you would *actually* do if you were caught out in fog (apart from the VHF calls you would make to the harbourmaster and other vessels).

4

In good visibility, one person (the exercise skipper for this purpose) is given full access to all the instruments including radar, but is not allowed to see outside the vessel (e.g. they go below, if that's where the instruments are located). The on-deck crew follow the exercise skipper's instructions: where to steer, how fast to go, what to look out for. The actual skipper is also on-deck, and is responsible for keeping the situation safe and not confusing other vessels too much, notwithstanding the instructions received from below!

The exercise task is very simple: navigate the boat slowly and safely along a chosen path, avoiding dangers (e.g. into or out of harbour, avoiding large ship channels).

4

First practise this with no other vessels around: navigate with GPS, chart plotter, chart, etc, with the radar assisting your navigation.

Then practise in a situation where there *are* other vessels around, and do what you would do in restricted visibility, using the radar for collision avoidance.

Stop and debrief whenever the on-deck skipper thinks a dangerous situation may be developing, or the exercise skipper has missed something.

Try to include a situation where you can do a radar plot, and compare your radar plot observations with the true situation – you might use a cooperating boat as a target for this part of the exercise.

Points to note

- Collision warning and avoidance is probably the principal use for radar

- **BUT** *collisions still occur, despite the widespread use of radar!*

- Main problems are: missing small targets, and becoming confused about what other vessels are doing.

- Avoid fog if at all possible.

- Use radar regularly to assist and complement visual look-out.

- Holding a steady course, use the electronic bearing line (EBL) to see if the target's bearing is changing.

- The IRPCS have different rules for vessels 'in sight of one another' or vessels 'in restricted visibility'.

- IRPCS is emphatic about proper use of radar, and systematic observation of detected objects.

- Target plotting (manual or automatic) can give you important information about the other vessel and aid your decision making.

- If in doubt, call on VHF.

- AIS can assist significantly, but is not a complete solution because not all vessels are transmitting (or receiving).

4

5 How to use radar for navigation and pilotage

Introduction

5

Our main concerns with both **navigation** and **pilotage** are to know where we are and to navigate in safe water. The advantage of radar is that it can help us to do this in conditions of low visibility.

Navigation, even in low visibility, has become a very much less demanding task due to the general availability of GPS, and even more so with the added aid of a chart plotter. However, as experienced skippers know, there is a sort of 'Golden Rule' of navigation, which is never to rely completely and solely on one source of information. It is all too easy for equipment to malfunction or mislead you, or just for mistakes to be made, and the way that you detect such occurrences is to arrange things so that you can cross-check information from (at least) two different sources.

This is particularly true when you are close to dangers. **Pilotage** is the process of navigating in relatively close quarters to land, so that observation of the land and landmarks becomes more important and relevant for keeping the vessel safe, and blind trust in the GPS alone is particularly unwise.

In good visibility we do this more or less continuously by informally 'seeing where we are' and, more accurately, by taking bearings on important landmarks. Measuring the depth is another form of cross-checking and, typically, a pilotage plan will include these techniques and more (transits, clearing lines, etc).

In poor visibility, or at night, radar becomes extremely useful for this kind of observation. Radar's usefulness is that it can assist in observing both the land and navigation aids, the observations becoming more accurate and useful as you approach relatively close to the objects in question. The radar is genuinely independent of your GPS and chart plotter (even though it may be interfaced to both). As explained in Chapter 2, the difference between them is that information on a chart plotter comes from a database with the input of your own GPS position, whereas everything in a radar picture is actually sensed by the instrument on the boat. (Note, however, that information such as the latitude and longitude of the cursor or a target *is* dependent on the GPS: this information is not detected by the radar but is generated by GPS input to the radar processor. Chapter 7 explains.)

Of course, radar cannot tell you about underwater dangers or anything that it cannot see, so your primary navigational tool is the chart, whether paper or electronic. What radar can do is:

■ detect land (but probably not low sand or mud), and give a general picture of the land which is in radar range and not hidden;

■ detect navigation marks (buoys, LANBYs);

■ identify **racons** (radar beacons);

■ accurately measure the distance to detected objects;

■ measure the bearing to detected objects
(in practice, less accurately than a hand-bearing compass);

■ from the above, confirm what the chart or chart plotter is telling you, and allow you to construct position lines and monitor clearing lines.

It will become obvious from the following sections that, useful though radar is, on its own it is not an ideal tool for navigation or pilotage. Use of radar alone for these purposes is not advised and, in the 'nightmare scenario' of your GPS being unavailable *and* thick fog, radar should be regarded as an emergency tool and used only with very great caution.

Which type of display?

It helps considerably for navigation and pilotage if you can use north-up mode on the radar display (if it's available), so that the radar picture is the same orientation as the chart. For one thing, it makes it much easier to switch from looking at the chart or pilotage plan to looking at the radar display. Secondly, it makes it much less confusing if you are changing course or heading quite frequently, because the radar display won't be changing every time you turn; with close-quarters pilotage, in particular, you tend to change heading and course quite a lot, both to avoid other vessels and to get onto your desired track.

The **true motion ground stabilized** display can also be useful as it keeps the display 'fixed' to the ground, and shows your own vessel's actual motion over the ground – in some ways a more natural presentation for navigation purposes. Details are given in Chapter 7.

Comparison between chart and radar picture

It is important not to expect a radar picture to give you everything that a chart does. Apart from the obvious (no names or underwater information), you have to be very aware of what targets the radar is capable of detecting – in particular:

■ they have to be good radar targets;

■ they mustn't be hidden, on a 'line of sight' from the scanner;

■ they must be above the radar's horizon.

5

Good radar targets are objects or features that are likely to scatter a lot of microwaves back towards the scanner, e.g.:

- a group of buildings;

- a group of metal structures;

- land with surfaces angled towards the scanner, e.g. rocky shores.

5

Because of the imperfections of the radar picture – i.e. some objects just don't show up on the picture – it can often be quite difficult to recognise and identify the features that do appear. Very often the *shape* of the coastline, which we instinctively use for reference, is incomplete, and doesn't look like the chart.

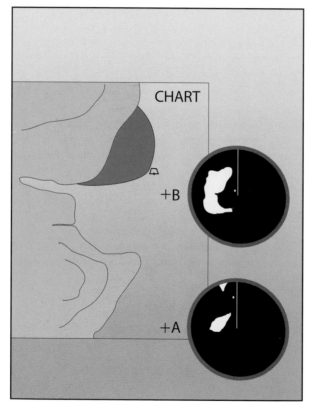

Figure 5.1 *Radar pictures from two different positions. Pictures are different because the high ground obscures the land behind it.*

Figure 5.1 gives an illustration of this. It shows the charted area, and the likely radar observations from vessels at two different positions. The points to note are:

- the pictures are different – so if you go from A to B, your radar picture will gradually change;

- the main reason for the different pictures is that, from A's position, high land in the foreground completely hides the land behind it;

- neither radar detects the very poor radar target: the mud on the northern edge of the estuary (fortunately, the port-hand lateral buoy has a radar reflector on it!);

- B's picture shows that very narrow targets, or gaps in a target, like the small river mouth to the south of the estuary, cannot be discriminated because of the relatively wide beamwidth – it's like trying to find a small hole by poking at it with a thick stick! For the same reason, B's picture doesn't show the full extent of the larger estuary.

Do not assume that just because something is a good navigation mark it is a good radar target. Lighthouses with circular walls are not naturally a good shape to reflect microwaves! Visually conspicuous landmarks may be indistinguishable from the surrounding land – and something that looks conspicuous on the radar display may *not* be obvious visually. Navigation buoys with radar reflectors will be about as visible as a small boat with a radar reflector.

The best way to learn what your radar can 'see' is to use it frequently – especially going in and out of your home port, where you can get accustomed to the various landmarks that the radar picks out well.

Identifying radar targets

5

Some modern radars, particularly those integrated with chart plotters, have the facility to compare the position of a radar target with the chart quickly and accurately – e.g. you move your cursor to the target, and its corresponding position is shown on the chart plotter screen, or you actually overlay the radar picture onto the chart display. This is obviously very useful – an undreamed-of luxury for an older generation of radar observers. You can, for example, tell whether a target is a charted navigation buoy or some other object (a vessel, for example), and you can quickly learn which charted features show up as radar targets.

In the absence of this, you have to do yourself what the integrated system would be doing for you. It helps if you can use north-up mode on the radar display, so that the radar picture is the same orientation as the chart. The method of identifying a specific target is as follows:

- measure the range and bearing to the target;

- if you are *not* using north-up, convert the measured (relative) bearing to true bearing. That is, add the relative bearing to the boat's heading (for ship's heading-up) or course (for course-up), correcting for magnetic variation and deviation in the normal way;

- plot this from the boat's position on the chart.

Of course in practice, once you have positively identified a few land targets, you can fill in the gaps, noting the probable identity of other targets from their position relative to the ones already identified. But even when identifying landmarks by eye, it is sometimes important to confirm their identity by taking a bearing. In the same way, you should always use a methodical approach when identifying radar targets: 'never assume' is always a good rule when skippering a boat!

Landfall

It's always a good moment when you get your first sight of land after a crossing. In less than perfect visibility, or at night, the radar may be able to spot it before you can with the binoculars.

There are several points to note about this. Firstly, you are trying to spot something at the radar's maximum range (and it may or may not be a good radar target). The strength of the radar returns diminishes very rapidly with distance (this is covered further in Chapter 6) and at maximum range you are pushing the limits of what the receiver can detect. Try turning the gain up while looking at the area of screen where you expect the land echoes, to see if this helps you to spot it.

Secondly, at this distance much of the land will be below the horizon – particularly a low coastline. You have probably noticed that the first things you spot by eye are hills some distance inland, with the coast appearing when you get much closer: the same applies to radar. This means the first targets spotted will generally be more distant than the boat's actual distance to the shore (and conversely, you are likely to be closer to the shore than the radar picture would lead you to believe).

Thirdly, the radar beam will be quite wide at maximum range – so any detected target will be displayed as wide as the beam, i.e. given a considerable 'spread' as the beam sweeps round.

So don't expect to recognize the shape of the shoreline – because the shoreline is probably not what the radar is detecting. Although you may *detect* land, it can be quite difficult to *identify* it, unless you pay careful attention to the charted topology of the land, what the radar is likely to detect, and the distorting effects of the beam width. And be aware that there may be land between you and the detected target – don't measure the distance to the nearest return and assume that is your distance to the shore.

Radar position lines

Just as you can get a **position line** by taking a bearing on a charted landmark, so radar can give you position lines if you can identify on the chart something that you can also identify on the radar picture. In fact, two different kinds of position line can be constructed from observing a radar target: one from its *range*, and the other from its *bearing*.

What are position lines and clearing lines?

A **position line** is a line drawn on a chart in such a way that *you know your boat is somewhere on that line*. The most common example is when you take a bearing on a landmark with a hand-bearing compass; you convert the bearing from magnetic to true, and then draw a line on the chart in the direction of the true bearing that runs through the charted position of the landmark.

Depending on how you get them, position lines can be straight (like the example above), curved (e.g. a light rising and dipping, which gives you the distance from the light), or even highly irregular (like a depth contour).

If you get two position lines then, logically, where they cross is your actual position. Because they might not be dead accurate, you get a better result if they cross at a good angle, preferably 60–90 degrees (crossing at a narrow angle will give you a *large* error in position for a *small* error in one or both of the lines).

The other thing is, you might have made a mistake. To guard against this, when you are fixing your position (and follow the 'Golden Rule' of always checking things against each other), you should generally plot a third position line to see if it 'agrees' with the other two. If it does not, then the wise navigator will try to discover where the mistake lies, rather than just rub out the line that makes the position fix look bad!

A **clearing line** is a line drawn on the chart such that: a) you have safe water to one side of it and dangers on the other; and b) you can quickly tell which side of it you are on. For the latter reason it is generally constructed as you would construct a position line, e.g. on a given bearing to a landmark. Almost any position line technique can be used to construct a clearing line.

Plotting radar position lines

The first step is to select a suitable discrete radar target, which can be positively identified and is marked on the chart. Place the variable range marker (VRM) and electronic bearing line (EBL) on the target, to get an accurate reading of its range and bearing (this is quite often a single action with the cursor on modern radar displays).

Set a pair of compasses to the measured range. Then with the point on the charted position of the target, draw the arc of a circle through the area of your likely location – this is the position line. This will be much more accurate than any visual estimation of the range.

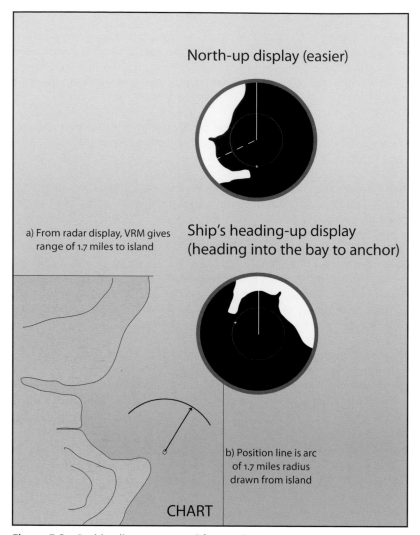

North-up display (easier)

a) From radar display, VRM gives range of 1.7 miles to island

Ship's heading-up display (heading into the bay to anchor)

b) Position line is arc of 1.7 miles radius drawn from island

CHART

Figure 5.2 *Position line constructed from radar range to a discrete target.*

The bearing, on the other hand, is likely to be *less* accurate than a hand-bearing compass; the beam width makes bearing measurement inherently less precise, and the method also depends on an exact measurement of the boat's heading when the target's bearing is measured, so this can be a source of error.

The technique with a basic ship's heading-up display is to measure the relative bearing of the target, and simultaneously note the boat's heading. Correct the latter for magnetic variation and deviation, and then add or subtract the measured bearing to port or starboard of the bow or ship's heading (add for a target on starboard, subtract for port). With a north-up display, the process is easier. Just take the bearing direct from the screen, but *do check* the means by which the radar is measuring ship's heading – which it needs in order to generate the north-up display. If, for example, it is actually using GPS Track (or Course Over Ground), this will probably *not* be the same as the instantaneous ship's heading and will introduce a further error; similarly, if a gyro compass or flux-gate compass is used, check whether there is any known deviation, and whether 'north-up' means magnetic or true. If it is magnetic north, you will have to correct the bearing for magnetic variation (just as with a hand-held compass bearing).

On the chart, draw a line in the direction of the true bearing so that it passes through your approximate position and through the charted position of the target (just as you do with a normal three-bearing fix).

5

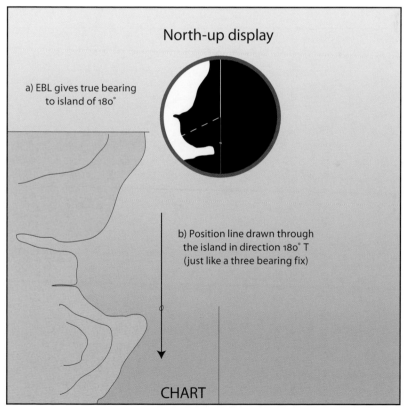

Figure 5.3 *Position line constructed from radar bearing to a discrete target: with north-up display*

Position fix using radar

Given the fact that the two types of radar position line naturally cross at right angles, it would be tempting to assume that range and bearing from a single object would make an easy position fix. In fact, this is not a very good plan. Firstly, as explained, the bearing position line is not particularly accurate. Secondly, there is a danger that you may not have identified the object correctly. This is a danger even with visual fixes, but it is far more difficult to be sure of the identification of a radar target – particularly in poor visibility.

The ideal situation is to use the radar to measure range (which it is very good at) and the hand-bearing compass to get bearings. This can give you two position lines, at right-angles to one another, from the same object – observed and identified both visually and by radar. The fix should then be checked against a third position line of either type *from a different object*, preferably identified by sight.

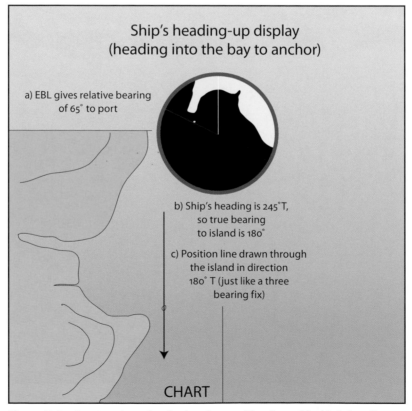

Figure 5.4 *Construction of radar bearing position line with ship's heading-up display*

The safest radar-only position fix is to use three ranges from three different objects. As stated above, range position lines are more accurate. The three arcs should cross nearly at the same point, with a small 'cocked hat' between them: if one of the objects has been incorrectly identified, or a range incorrectly plotted, it should be obvious that the 'cocked hat' is excessively large (just like a normal three bearing fix).

Racons (radar beacons)

So far, I have discussed the fact that navigation aids do not lend themselves to detection, let alone recognition, by radar nearly as well as they do by eye. Clearly, radar will not pick up a lighthouse or buoy just because it is lit, and certainly won't tell us its light characteristics. However, there is an important and useful exception to this.

A **racon** – radar transponder beacon – detects when it is being scanned by a ship's radar and very quickly transmits a signal on the same radar frequency. This is picked up by the ship's radar receiver, and displayed on the screen (to the receiver, it is indistinguishable from a strong radar echo)

The pattern displayed will be a distinctive 'flash' extending outwards from the position of the racon. In some cases, this is broken into the dots and dashes of a morse code letter which is stated on the chart, enabling you to positively identify this navigation aid. Figure 5.5 below shows an example.

Note that the target itself is just closer than the closest point of the transponder-generated flash. Sometimes, two racons are used to define a leading line into port. The flashes line up on the display when the ship is on the leading line.

Note also that, because the racon has to cater for different ships whose radars are not all tuned to exactly the same frequency, you may not see it every time your radar scans it. Depending on the racon design, you may see it two or three times in a row and then not see it again for a minute or so.

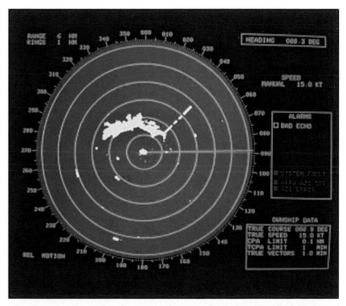

Figure 5.5 *Example of a racon detected by radar. The 'flash' beyond the target is morse code X-ray (– · · –)*

5

What is a SART?

The other kind of transponder you should be aware of is a **Search And Rescue Transponder**, or **SART**. This device is carried in some liferafts, and is one of the Distress Signals defined in Annex IV of the IRPCS.

When it detects that it is being scanned by radar, it transmits a series of 12 short pulses which are then displayed on the scanning radar's screen. At a distance these will be seen as 12 dots, and as you get closer they will expand into arcs of increasing size and may become full circles at close range.

As it is a distress signal, if you observe one you should report it to the rescue authorities (e.g. UK Coastguard) and keep a good lookout for any sign of a liferaft or vessel in distress.

Figure 5.6 *A Search and Rescue Transponder (SART)*

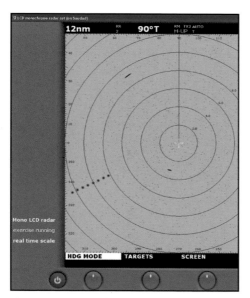

Figure 5.7 *SART detected by radar. Left: SART at approx 8 mile range, with dots showing. Right: SART at approx 3 mile range, with arcs showing. Note that you may not see all 12 dots or arcs (depends on the range scale in use).*

Useful techniques

Given what I have said above regarding position lines and clearing lines, it is possible to devise radar pilotage techniques for proceeding along a preplanned safe track, guided by the radar picture.

On the whole, these are only recommended for experienced operators using high specification radar equipment. For example, if you have a reliable, large, well-stabilized north-up display together the facility to plot lines on it (electronically generated is preferable to Chinagraph pencil), then you can define a path along which you want a particular radar target (of a fixed navigation aid or landmark) to travel – and thus ensure that your boat is proceeding along a predefined safe track. This is known as *parallel indexing* and because it is a relatively complex procedure, requiring facilities that not all small craft radars are likely to possess, I am not going to describe it in detail (see References at the end of the book).

What is more likely is that you will want to use your radar to *check* your pilotage, and a particularly useful attribute is its ability to measure distance off a suitable shore target.

Say, for example, you are piloting out of Plymouth harbour towards the Yealm River (see chart in Figure 5.8). Having come out of the Eastern Channel out of Plymouth Sound, your pilotage plan is to pass Shag Stone by going midway between its spar (unlit) and East Tinker east cardinal buoy (you can see both

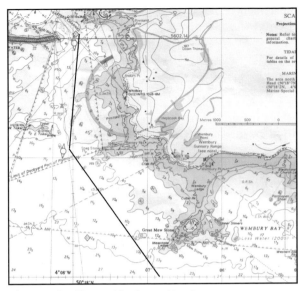

Figure 5.8 *Chart with planned track*

5

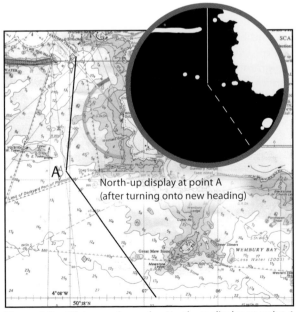

North-up display at point A
(after turning onto new heading)

Figure 5.9 *Pilotage using radar. North-up display at point A shows the land, breakwater and two navigation buoys.*

those marks). You then want to round Great Mew Stone keeping a safe distance off Mewstone Ledge, but it is not immediately obvious how to make sure that you are a sufficient distance off. There are no obvious leading lines or contours to follow – and the line of yellow buoys that used to be there for firing practice have been removed!

One possible technique would be to set a course (measured from the chart) to clear the ledge at a suitable distance, and then monitor Great Mew Stone (not a bad radar target) on the radar screen as you approach the Ledge. If you are using the basic ship's heading-up or azimuth stabilized course-up display, then the target (which is ahead of you on port side) should come vertically down the screen, provided you are proceeding in the direction of your heading (or course) – i.e. you have little or no tide or leeway setting you off your course. This is a relatively straightforward thing to visualize on the display, so perhaps this is one instance

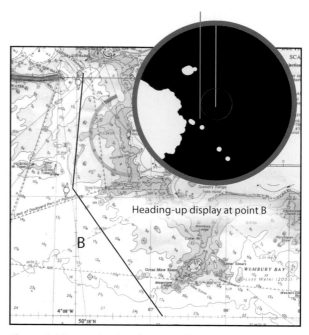

Heading-up display at point B

Figure 5.10 *Monitoring distance off the Mew Stone target. As long as it tracks down the display to the left of the blue line, the required distance off (marked by the magenta VRM circle) will be maintained.*

when using the ship's heading-up, or course-up, display is not such a bad thing for pilotage, and you can easily monitor other vessels for collision avoidance at the same time.

If you are using north-up display then the target should move towards you parallel to your heading or course line.

Set your VRM to the required distance off (you are effectively setting up a clearing line by doing this: safe if the target land is outside the circle, dangerous otherwise). Using a straight-edge on the screen, you can quickly see whether the target will pass inside or outside the VRM minimum range circle when you pass it. If it starts to come too close – possibly because you are tracking further to the east than intended (tide, leeway) – then alter course to starboard until you are happy that as the Mew Stone target tracks down the screen it will pass *outside* the VRM clearing line circle.

The strength of this plan is that the radar is helping you to monitor precisely the thing that concerns you – the distance off the land.

Practical exercises

1 Position lines and position fixing

Take any opportunity to practise identifying land feature radar targets. You can learn for yourself which targets show up best, and hopefully work out why that is from the theoretical explanations in this book and your own experience.

With the boat stationary a safe distance offshore, construct position lines using:

- radar range to an identified target;

- radar bearing to an identified target.

Construct some 'radar only' position fixes.

From the quality of the fixes (the size of the triangle or 'cocked hat') and comparison with visual fixes and GPS, note the probable accuracy of both range and bearing position lines. Confirm that range lines are generally more accurate than bearing lines.

The safest 'radar only' position fix is three ranges from three different objects: practise this.

Practise a combined 'hand bearing and radar range' position fix off a single object, checking against GPS or a third position line from a different object.

2 Pilotage using radar distance-off

In a pilotage situation, practise monitoring radar distance off whilst passing a radar-conspicuous headland, as described in the example in this chapter. Take care to correctly identify the radar target – e.g. do not assume that sand or mud extending from the headland will show up on the display!

5

Points to note

■ Radar is useful as a cross-check in navigation and pilotage – particularly in poor visibility or at night.

■ Radar will detect land and navigation buoys – but not everything 'hard' is a good radar target.

■ Interpreting the radar picture is not always straightforward and requires practice.

■ At long range you will not see the coastline below the horizon, and beamwidth makes the picture quite coarse and distorted.

■ Radar can be used for range position lines which are circular, centred on the observed object.

■ Bearing position lines can also be derived, but these are less accurate than a hand-held compass.

■ Racons and SARTS are both radar transponders – the former for navigation marks and the latter for distress.

■ Using radar distance-off is a useful pilotage technique – using radar on its own for pilotage is not advised.

6 More radar science – confusion and anomalies

Up until now, we have looked in just enough detail about how radar works to give you a basic knowledge and ability to interpret the picture. It's time to look at the science in more detail and to explain some of the confusion and anomalies you can get with a radar picture. If you want extra information about how radar works, or your radar picture just doesn't make sense, then this is the place to look for an explanation. However, this is not the book to give you a treatise on microwave physics or radar engineering: there are some references for that kind of thing at the end of the book.

Radar frequency bands

The electromagnetic spectrum is used for lots of different human activities, and so it is necessary to have international agreements about which frequencies are used for various purposes. Two groups of frequencies are allocated to civil marine radar and these are known as:

- **X-band:** frequency 9320–9500MHz: wavelength approx 3cm;

- **S-band:** frequency 2900–3100MHz: wavelength approx 10cm.

Within these bands, individual radars are tuned to different frequencies so that (hopefully) they don't interfere with one another.

Because of the minor differences in the way the higher and lower frequencies behave, the two types of radar are good at slightly different things. For example, the S-band can give better performance in clutter and rain; the X-band can give a stronger return for a given size of target. Big ships can have one of each, and may use either or both. Most small boat radars, however, are X-band; the shorter wavelength requires a smaller scanner to achieve a suitably narrow beam width, which is obviously an advantage on a small vessel.

Signal strength and range

In order to make you more generally aware of what's going on, it's worth appreciating how the strength of the returned signal decreases with the range of the target – and what the radar has to do about this.

If you have a single source of radiation – a light bulb in an open space, for example – the brightness of the light decreases as you move away from it, in proportion to the *square* of the distance.

Therefore, if you go from 1 mile to 10 miles, the brightness you can see decreases by a factor of 10^2 (which is 10 x 10), or 100.

With radar you have two such effects: the 'brightness' of the radar signal you are sending out, and the 'brightness' of the reflected signal that comes back to the receiver. The result is that the signal strength, for a given target, decreases in proportion to the *fourth power* of the distance.

Say you have two identical radar target objects: one of them is 1 mile away, and the other 10 miles away. The signal from the nearer one is 10^4 times (i.e. 10,000 times) as strong as the signal from the further one. That makes the further one quite a lot harder to detect. (Given that, in addition, by no means all the energy is reflected back from the target, it is not surprising that radar needs a lot of power and has limited range.)

This also means that radar has to do quite a nifty job when it comes to displaying the picture, because the observer wants to see similar targets at different ranges displayed at similar brightness. What radar does is *increase the gain* of the receiver in the right proportion according to the distance – amplifying signals received from 10 miles by 10,000 compared to signals received from 1 mile. So it is not surprising that the picture quality may not be as good at extreme ranges, because some of what is amplified will be *noise*.

Pulse length and pulse repetition frequency

In Chapter 2 I explained that, in order to measure range, the radar sends out a short pulse of microwave energy and then stops and listens for echoes. This process is then repeated a short time later. The duration of a pulse is called the **pulse length**, and the number transmitted per second is the **pulse repetition frequency (PRF)**. On most radars these will be set automatically, but on some the user can influence the selection of pulse length to enhance the display of targets (see section on **target expansion** below).

If you look at the technical bit of your user manual, you may find a table which tells you the pulse length (in microseconds) and PRF (in Hz, meaning 'per second') that your radar uses, and these depend mainly on the range scale selected (plus any other selection that may be available to the user):

- short range scale: short pulse length and high PRF;

- longer range scale: longer pulse length and lower PRF.

Why?

PRF is easiest to explain. The radar sends out a pulse, waits until all the possible echoes have been detected, then does it again. If it sent another pulse out too quickly, the echoes from the first one would become confused with those from the second. Obviously, how long you have to wait depends on how far you expect the echoes to come back from: hence higher PRF (shorter wait) for shorter range scale settings.

Pulse length has two effects. A longer burst of microwave radiation contains more energy, so echoes – particularly weak, distant echoes – are easier to detect. On the other hand, when you detect the echo, a very short pulse enables you to calculate its range very precisely, and a longer one will spread the target in a radial direction and can even obscure and merge two echoes from two targets on the same bearing at slightly different ranges.

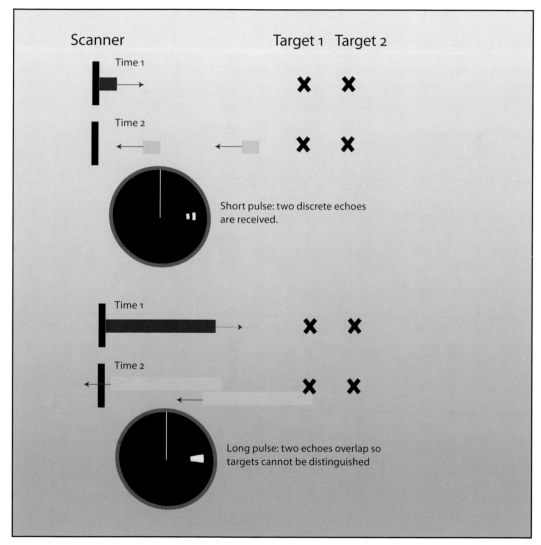

6

Figure 6.1 *This shows how a short pulse length can produce better range discrimination, and a long pulse length can make two targets (at same bearing and different ranges) merge together.*

Say you have a pulse length of 1 microsecond (the longest on my radar). You can imagine that as a train of microwaves 300m long (that's the pulse duration multiplied by the speed of light). Any echo received back will be 300m longer than the theoretical minimum from a very short pulse. This is divided by 2 to work out the range of the target, because the total distance travelled by the pulse is out and back. So every target will be displayed 150m 'thicker' in range than it actually is. Furthermore, two targets up to 150m apart in range and on the same bearing are likely to touch or merge together, so you will not be able to discriminate between them.

Clearly this matters little if you are looking at targets 10 miles away, but for ranges of half a mile (1000m) or less it would give a very coarse picture indeed – hence shorter pulse length is used at these ranges (where, in any case, you don't need as much energy because the reflected signal strength is very much greater). Good **range discrimination** is useful if, say, you are plotting targets for collision avoidance; using a shorter range scale setting will improve this.

Target expansion

Some radars offer the facility to increase the pulse length slightly, thereby making the targets larger and more visible, but at the same time losing some range discrimination.

Horizontal beam width

I've already mentioned beam width in previous chapters, because it is important to understand its effect when interpreting the picture and **bearing discrimination** of targets. As the radar scans round, any target will be displayed for as long as it remains in the beam, so if the beam is relatively wide the target will seem wider than it

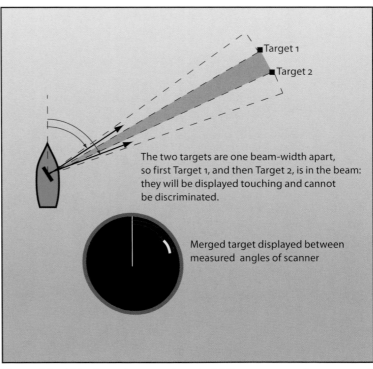

Target 1
Target 2

The two targets are one beam-width apart, so first Target 1, and then Target 2, is in the beam: they will be displayed touching and cannot be discriminated.

Merged target displayed between measured angles of scanner

Figure 6.2 *This shows how two targets (at the same range) can merge if their respective bearings are closer than one beam width. So narrow beam width gives better bearing discrimination.*

actually is – 'stretched' in the tangential direction. By the same token, the target's actual bearing, and size in the tangential direction, will be more accurately represented if the beam is narrow.

If two targets are close together, at the same range but slightly different bearings, their returns will merge together on the picture to look like a single target.

To see what this means in practice, let's consider two radars: a commercial standard big ship radar with 2° beam width (some ship radars have less than 1°) and a medium quality small boat radar with 5° beam width. You can get an immediate (and very relevant) idea of these beam widths by looking at your hand-bearing compass. We are used to getting 1° precision when measuring bearings with the compass (or trying to), so as you look at distant vessels and coastline, you can visualize how relatively coarse the picture is going to be if it cannot discriminate the bearing to better than 5°.

Another way of looking at it is: how wide is the beam at various ranges? The answer is, at 1 mile the 2° beam is 65 metres wide and the 5° beam is 160 metres wide. At 10 miles the narrow beam is over a third of a mile wide, and the 5° beam is nearly 0.9 miles wide.

Of course these can be perfectly acceptable: in fact, a small LCD display may not be capable of showing better angular precision than that. If your principle use for the radar is to detect other vessels, then it is no bad thing if these are shown on the display as big blobby targets which you can see easily at a glance. It matters little if two targets a few hundred yards apart look like one when they are 2 miles away.

You may well notice that when you steer the boat to pass close to a distant headland, the radar picture will *not* necessarily show the headland with the heading line on its extremity (where it 'ought to be'): instead the heading line will pass through the end of the headland on the display. This is the effect of beam width.

Although narrow beam width is a 'good thing' for angular discrimination and presentation of targets, there are two trade-offs in the radar design:

- in order to get a narrower beam width you will need a wider scanner, and consequently more bulk and weight up the mast;

- it is easier to get more power into a wider beam, and this can be an advantage in picking out poor radar targets, and at longer range.

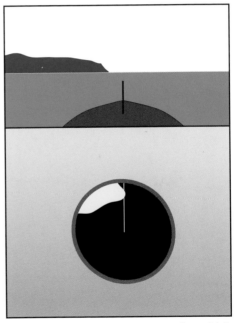

Figure 6.3 *Radar picture may not show ship's heading line 'clearing' headland, because the target (the land) is 'spread' by the effect of beam width.*

Vertical beam width

This is a different matter. As our radar scans around, we want it to see 'everything' as it looks in a particular direction, i.e. down far enough to pick out a small target at sea level close to our boat, and up far enough to see high bits of land. We want it to do this even when the boat is rolling, heeling or in the case of a motor boat, at planning attitude. A typical **vertical beam width** is 25° for a small boat radar.

This should have very little effect on the radar picture, but is worth bearing in mind if your small boat is in a very rough sea. Heeling (on sailing boats), combined with rolling and pitching, will 'twist' the beam, effectively giving a wider horizontal beam by virtue of the relatively large vertical beam width. Extreme attitudes may cause the beam to point down into the sea or up into the air, thereby missing targets at close range. Of course, in these conditions you may have other things than the quality of your radar picture to worry about…

Side lobes

The ideal is to engineer a single, clean narrow beam, but this is not quite possible due to the way the antenna operates. Other, less powerful beams are generated at an angle to the main one. These are called **side lobes**, and they can cause some confusing effects.

If you have a strong radar target at close range, in addition to the return from the main beam the receiver can pick up echoes from the less powerful side beams as well. But it won't know these are from the side lobes, so it will display extra (false) targets in the direction the main scanner is pointing in when it receives the extra echoes.

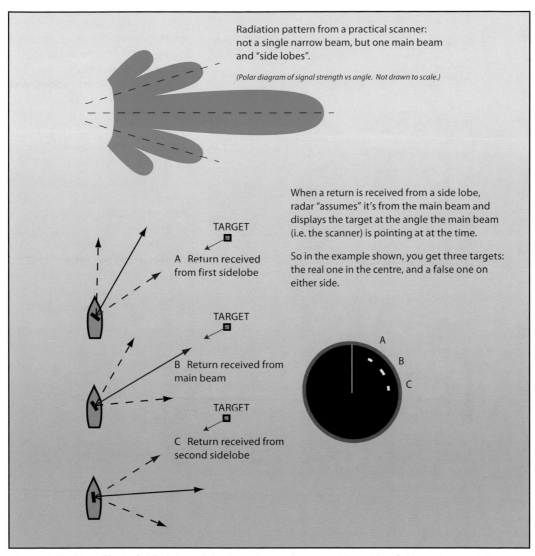

Radiation pattern from a practical scanner: not a single narrow beam, but one main beam and "side lobes".

(Polar diagram of signal strength vs angle. Not drawn to scale.)

When a return is received from a side lobe, radar "assumes" it's from the main beam and displays the target at the angle the main beam (i.e. the scanner) is pointing at at the time.

So in the example shown, you get three targets: the real one in the centre, and a false one on either side.

TARGET

A Return received from first sidelobe

TARGET

B Return received from main beam

TARGET

C Return received from second sidelobe

A
B
C

Figure 6.4 *The effect of side lobes. False targets may be seen either side of the real one, at the same range. Normally this is only a problem at close range.*

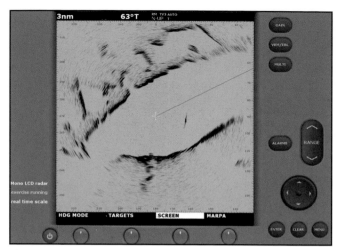

Figure 6.5 *Side lobe effect. Note the elongation on either side of the real target.*

Figure 6.5 shows what this effect looks like: additional weak targets appear on either side of the main strong one, at the same range, or a continuous arc may appear. This only occurs at short ranges, and the effect can be reduced using the sea clutter control.

Indirect and multiple echoes

We know that radar signals can reflect off objects, so it should come as no surprise that some of the returns detected by the receiver have come by a slightly longer route than the standard one (a single reflection off one object), and have in fact been reflected off several objects.

On a ship, you can get secondary reflections off deck cargo or superstructure – but this is unlikely to be a problem on a small boat with a good radar installation. Other, more likely, effects are:

■ Figure 6.6 shows what happens when you get multiple reflections between a target ship and your own boat – effectively, these appear as additional, false targets at a greater range than the true one, the give-away being that they are spaced out at equal intervals. Again, this is not very likely to be a problem on a small boat (which isn't very good at reflecting radar anyway) or at long ranges.

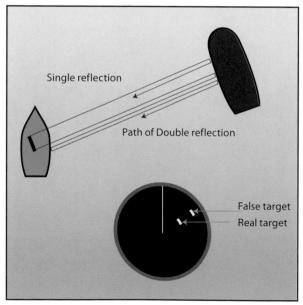

Figure 6.6 *False target caused by double reflection between own vessel and target.*

■ Figure 6.7 shows what happens if the return from a target bounces off another object like a harbour wall. A false target will appear exactly like a reflected image in a mirror (and the 'mirror' object itself should appear as a strong target).

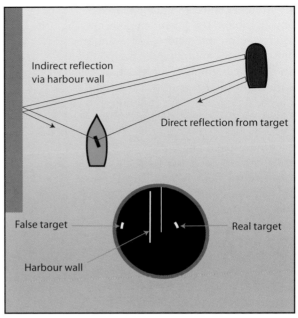

Figure 6.7 *False target caused by 'mirror' reflection off a harbour wall.*

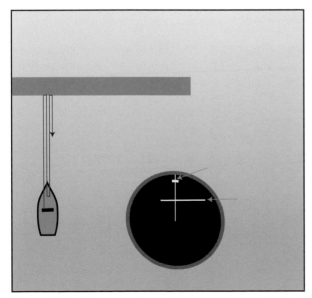

Figure 6.8 *False target caused by your own 'mirror' reflection in a bridge ahead of your vessel*

■ Figure 6.8 shows a particularly worrying situation where you get your own boat 'reflected' in a mirror object (like a bridge) ahead of you. It's worrying because it looks like another vessel approaching on the far side of the bridge, and when you manoeuvre to avoid it, it moves too, so that you are still on a collision course!

■ Another confusing situation involves power cables. The power cable is not itself seen on the picture, but a single 'false' target can appear at the point on the cable where it intersects with a perpendicular line from the boat – which, again, can be confusing in a waterway. Figure 6.9 illustrates this.

6

It can be seen that many of these are likely to be situations close to shore (with false targets sometimes on the land) where, in a small boat, you are unlikely to be using radar or placing any great reliance on it; but nevertheless it is worth being aware of the effects.

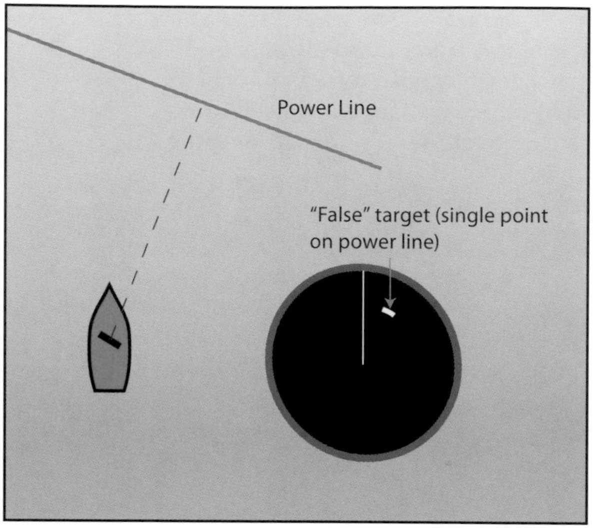

Figure 6.9 *The effect of a power line.*

Interference from other radars

Where other radars are operating in the same area, some may be on frequencies close to your own radar's frequency. This can cause interference, because their received pulses will be interpreted by your radar as echo returns. Generally the false returns will appear as a pattern on the radar display, so that you will know something is up. It is most likely to happen on long range scale settings.

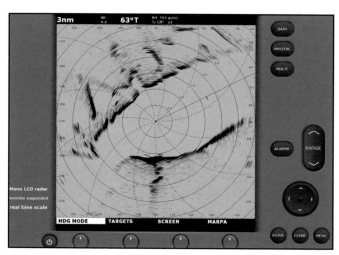

Figure 6.10 *Interference from another radar causing curved 'spokes' radiating from the centre of the radar picture.*

The radar will normally provide a facility to eliminate this interference – which may be 'on' as the default selection – so the chances are you may never see this effect.

Refraction and range

I mentioned in Chapter 2 that microwaves travel in straight lines, and that therefore you cannot see beyond the horizon. This is *almost* true!

What actually happens is that the atmosphere acts as a lens, and the path of the radar beam does, in fact, bend round the curve of the earth to a slight extent. The formulae given in Chapter 2 refer to *standard* atmospheric conditions; the *actual* atmospheric conditions, which vary from day to day and from place to place, may bend the beam slightly more or slightly less.

It's not necessary for us to predict this, but rather to be aware of the possibility that the range at which the radar can detect over-the-horizon objects can vary. This is similar to what you may have observed with VHF radio propagation – when you are well beyond 'line of sight' range from the English coast – off Brittany, say – you can sometimes hear UK Coastguard transmissions.

Points to note

- Civil marine radars use two frequency bands: X-band and S-band. Most small radars use X-band, and individual radars use different frequencies within the band.

- Interference from other radars can usually be eliminated automatically.

- Strength of returns from similar objects decreases in proportion to the fourth power of the object's distance.

- Pulse length varies automatically – longer for greater range.

- Pulse length can be increased manually to enhance targets – but this reduces range discrimination.

- Pulse repetition frequency is automatically reduced with range.

- Narrow horizontal beam width gives better angular discrimination, but requires a wider antenna and more power for the same strength of target return.

- Vertical beam width is much greater than horizontal.

- In addition to the central beam, radars have side lobes which can give rise to false targets either side of the real one at short range.

- False targets can also be generated by indirect returns – reflected from other objects.

- Atmospheric refraction can affect the radar horizon.

6

7 Modern radar – more advanced functions

Electronics manufacturers have been busy blurring the boundaries of the instruments that we carry on small boats and yachts. Also (as every mobile phone owner knows), they fill every device with as much functionality as they can think up! While this is admirable, and can be very useful, it can also be somewhat confusing. To some extent the world is divided into:

■ people who want to understand and use every clever function that is provided with every piece of kit they possess; and

■ people who want to go boating/sailing/diving/fishing!

Essentially, the author makes no apology for belonging to the second camp. I see my task in this book as follows:

■ to help the reader understand what radar is, and does – and that includes understanding what it *doesn't* do;

■ to explain the basic functions and how to use them to do the things radar is good at;

■ to explain the more advanced functions, many of which rely on integration with other instruments, in such a way that the user can understand and assess them, and see whether and when they will be useful.

The first two points are covered in Chapters 2 to 6. The last is the subject of this chapter. Ultimately it is your choice whether or not you use these functions, or consider them important when selecting a radar system to buy. This book's purpose is to explain what they do and where they can be useful – not to persuade you with my opinions!

'Basic' and 'advanced' functionality

'Stand-alone' radar is now a rarity. Most sets are integrated with other instruments and have gained greatly from that in terms of the functions they can offer on the display by combining or co-presenting information. What has *not* changed as much, however, is the basic radar performance, i.e. what the radar *itself* can do – the limitations described in Chapters 2 and 6 still apply (although the engineers have done some pretty nifty design work, such as allowing a radar to operate on two range scale settings at once).

It is quite important to understand the distinction between *radar performance* and *integration with other instruments* when choosing a radar system, and understanding what each of these can do for you.

In Chapters 2 and 3 I described the following basic radar functions:

- basic display (plan position indicator) with ship's heading-up;

- power on, standby and transmit;

- range scale;

- brilliance, or light and contrast;

- gain;

- tuning;

- azimuth stabilization – course-up, north-up;

- sea clutter;

- rain (precipitation) clutter;

- range rings and ship's heading marker;

- electronic bearing line;

- variable range marker.

These functions enable us to do pretty well everything described in Chapters 4 and 5.

In this chapter I describe and explain the following more advanced functions:

- look-ahead/zoom in;

- more on azimuth stabilized course-up and north-up display (already explained a bit in Chapter 2);

- target latitude/longitude;

- chart overlays and waypoints on the radar display;

- AIS integration;

- true motion – sea and ground stabilization;

7

- indications of signal strength: brightness, colour;

- dual range display;

- high speed update;

- scan-to-scan correlation;

- automated target tracking (MARPA/ATA);

- target wakes/echo trails;

- guard zones;

- watchman mode.

Once again, I can only describe these features in generalized terms. You will have to refer to the *Owner's Handbook* for your particular radar system if you want to see what functions are available, and how to operate them.

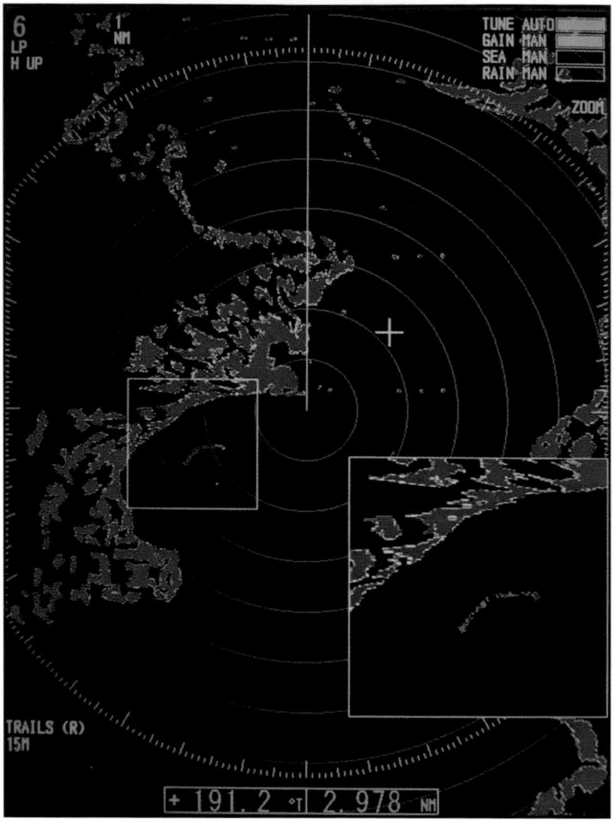

Look-ahead/zoom in

These functions will not seem particularly clever to anyone who is used to the way computers can

Figure 7.1 *User is able to 'zoom in' on the part of the picture, to enlarge a particular target.*

manipulate pictures that are being displayed on a screen – but just a few years ago they would have seemed like magic to the operator of a traditional radar system.

The picture data is still obtained in the same way, so the radar function is still exactly as described in Chapter 2, and no additional information is required from other instruments. However, once the picture is stored in the radar's processor, it can be manipulated and displayed in different ways on the screen. Thus you will be able to display the part of the picture which interests you – moved up or down the screen, and/or enlarged – using controls which are analogous to those of a drawing or photo package on a PC.

Normally, the bit which interests us most (particularly for fast-moving craft) is the area ahead of the boat, so **look-ahead** zooms in on that part of the picture for us. Just beware of looking at this area exclusively when keeping watch: there is always a possibility that something faster than you may be coming up from behind!

Azimuth stabilized display

Because this is a particularly important function for making the basic radar picture more useable, I described it briefly in Chapter 2.

The basic radar 'knows' the bearing of targets purely by where the scanner is pointing when it detects them. This is the physical angle between a ship's head reference point, and where the scanner is currently pointing during its 360° rotation. Any target returns are then displayed at this angle from the ship's heading line on the basic radar display.

This is fine if the boat is steady, but if the heading is changing – either because of a change of course or because of rough seas – the radar picture will rotate back and forth, making it difficult to see a constant and coherent picture and to identify small targets from one sweep to the next.

A solution to this is available if the radar processor knows, from one instant to the next, the actual direction the ship's head is pointing in. It can then calculate the absolute (rather than relative) direction of the scanner, and display a picture showing the absolute direction of the targets. This, of course, is

7

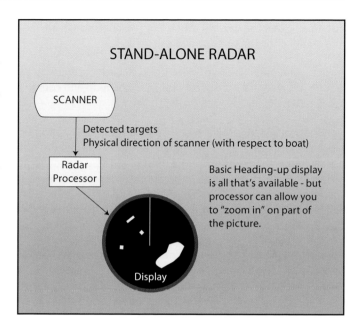

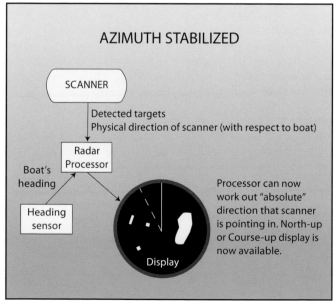

Figure 7.2 *Information block diagrams showing stand-alone radar (top) and azimuth stabilization (bottom)*

not affected by the yawing motion of the boat, so the resulting picture is **azimuth stabilized** and much easier for the operator to use.

From the designer's point of view, if you can achieve this you can theoretically display the picture any way up you want – but the most useful are **north-up** and **course-up**, and these are the ones provided on most systems.

North-up shows the radar picture in the same orientation as an ordinary chart, so it may be the preferred orientation for navigation and pilotage, where you need to relate the radar picture to a chart or pilot book sketch. The display will stay steady as you manoeuvre the boat by changing its heading – making it much less confusing, particularly in a pilotage situation where you might be changing heading quite a lot in order to get to your desired track. The ship's heading marker will normally be shown on the display, and of course it will not generally be pointing straight up the display from the centre as it is with a ship's heading-up display.

Course-up is in some ways the more natural presentation, and at first glance it seems the same as the basic display, only steadier. The course direction (12 o'clock on the display) is where you would *like* the boat to be heading. It can be taken either from the autopilot (the course you have selected), or from the boat's actual heading at a particular time, e.g. at the instant you select the display mode – useful if the boat is being steered manually. We then expect the boat's heading to vary slightly either side of the course direction, but the radar picture will be steady. An indication of ship's actual instantaneous heading (the ship's heading marker) will normally be shown on the display.

The picture then represents the same orientation as the basic ship's heading-up display, with targets displayed at approximately the same angle from the bow that the look-out observes – clearly an advantage for identifying targets on the display.

When you change course, however, you will need to set the display to the new course (unless this happens automatically from the autopilot). If you don't do this, the picture will stay at the previous orientation and this could be a source of considerable confusion if you assume you are still looking at a course-up display.

Ship's heading data

The performance and success of the azimuth stabilization function depends very much on the input of heading direction. Some electronic direction sensors are perfectly adequate for their primary function (normally autopilot), but may not provide heading data frequently enough to produce a good azimuth stabilized radar display.

A **fluxgate** compass is an electronic device that can sense in which direction it is pointing by sensing the earth's magnetic field. In this sense it is similar to a normal magnetic compass, except it senses the field (known to physicists as 'magnetic flux', hence the term fluxgate) and displays the direction information using a completely different technology. It is, incidentally, susceptible to the presence of ferrous materials near it, just like a magnetic compass. It is very accurate in direction, but it can be slow to respond with accurate data if the boat yaws rapidly.

More advanced technology combines the fluxgate compass with a **rate gyro**: the latter is good at sensing the *rate of change* of heading, but is unable to detect the absolute heading. If you combine the data from both sources you get what you need for a good azimuth stabilized radar display: accurate heading data which is highly responsive when the heading is changing quickly.

A better (and more expensive) heading sensor is available in the form of a **GPS compass** or **satellite compass**. GPS can sense position, not direction. The direction a single GPS can give you – **course over ground (COG)** – is simply taken from the last few 'remembered' positions it has been in, and is the direction in which they have been proceeding; so firstly it is historical, and secondly it has nothing to do with the direction the vessel is pointing in. If the vessel has been going sideways over the ground, that is the direction that COG will give you.

With two GPS receivers, however, you *can* get a direction. In a GPS compass they are placed a short distance apart and they give different positions: and thus a true direction between the two positions can be calculated. The calculation can be further improved if accelerometers are incorporated into the unit, to give improved data on the relative position of the GPS receivers when the vessel is rolling and pitching. These units are highly accurate, but rather bulky and need a clear view of the sky – so they are therefore harder to install. Currently, they are not particularly cheap. So although they can give excellent heading data, they are less likely to be suitable for small craft.

7

7

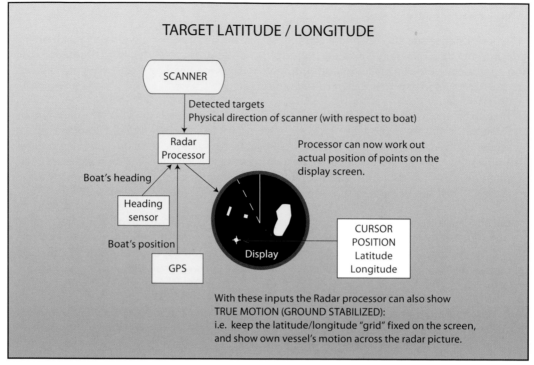

Figure 7.3 *Information block diagram showing how target latitude/longitude can be deduced: and how a True Motion Ground Stabilized display can be produced.*

Target latitude and longitude

In the previous section we saw how, with the addition of one piece of information (ship's heading) we can deduce the actual bearing of a target. With a further piece of information – our own position, supplied by GPS – the processor can work out the actual *position* of a target.

What normally happens is that you identify a radar target using the cursor, and the system will give you a readout of the cursor's position in latitude and longitude.

This function can be used for a number of tasks:

■ if you have a radar target which is another ship, you can determine its position if you want to call it on VHF ('Ship in position… this is…');

■ if you think it might be a navigation mark, you can confirm this with the mark's charted position;

■ if your radar is integrated with your chart plotter, positions (waypoints, target positions) can be transferred from one to the other (see next section); and finally

■ the function helps you to tie together an AIS message with a radar target, because the AIS message contains the sending ship's position. This can be done manually or (with integrated sets) automatically.

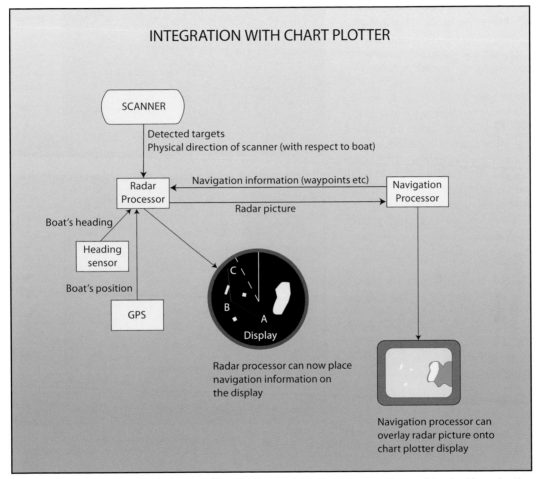

Figure 7.4 *Information block diagram showing how radar information can be combined with navigation information from the chart plotter.*

Combining radar with navigation information

Once the system knows the latitude/longitude position of all the points in your radar picture, it can integrate the radar picture with a chart plotter display and/or any other navigation data that your system contains. For example, it can superimpose waypoints, routes or tracks onto the radar picture.

Some systems allow you to overlay the radar picture onto the chart. This can be very useful, not least when you are inexperienced and learning to interpret the radar picture. All the explanation in earlier chapters about what will, and will not, be detected and displayed by the radar, will literally be demonstrated for you when you use this function.

Clearly it will help you to identify the land or navigation mark features that your radar is detecting, and also to pick up radar targets in the sea that are *not* shown on the chart, which are likely to be other vessels.

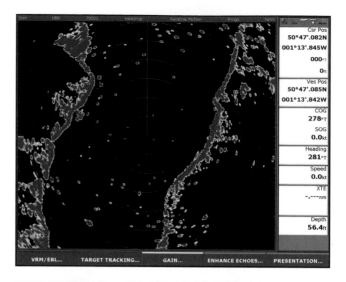

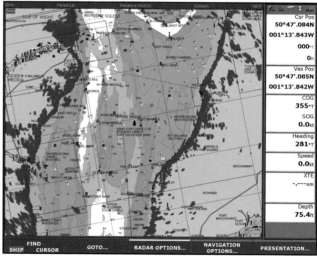

Figure 7.5 *Example of a radar picture overlaid onto a chart plotter display. Top: radar picture. Bottom: combined display.*

AIS integration

The functioning of an **automatic identification system (AIS)** is described in Chapter 4. From a technological point of view AIS has nothing to do with radar, but the two systems can usefully be integrated and that is why I mention it again in this chapter.

An AIS message can be associated with a specific radar target by means of the transmitting vessel's *position* (latitude and longitude), which is contained within the message. If you can obtain a readout of the cursor latitude and longitude (see above), then you can identify the radar target from the AIS message. This can also be done automatically, and indeed other information (course, speed) can be taken from the AIS message and displayed, numerically or graphically, on the radar display.

7

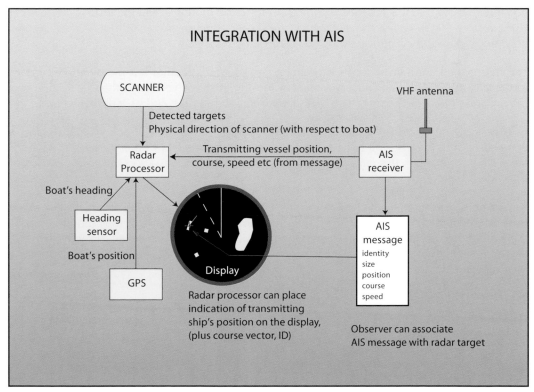

Figure 7.6 *Information block diagram showing how AIS information can be correlated with radar picture.*

True motion display: ground stabilized

In the basic ship's heading-up display, everything on the display is positioned as it is detected, i.e. at the observed range and bearing from the observing vessel, with the observing vessel at the centre of the display. The azimuth stabilized displays (course-up or north-up) are similar, with the observing vessel at the centre of the display. So the movement of targets on these displays is *relative to the observing vessel*: for example, land targets will move across the display at the speed your vessel is moving forward over the ground, and target vessels on a collision course will stay on a constant bearing as they approach. This kind of display is referred to as **relative motion**.

Given that you can work out the latitude and longitude of the targets on your radar picture, on some systems you can elect to keep the position grid fixed like a chart, and show your own boat's position moving across it (as, of course, you can with the chart display). This is known as **ground stabilization** (because the picture is 'fixed to the ground'), and the display is known as **true motion**. All the targets that are actually fixed (land, buoys) will be static on the display, and only moving objects will appear as moving targets.

This helps when using the radar for navigation and pilotage, because it is more natural to think of the land as stationary and your own boat as moving. It allows you to see the true motion over the ground of non-stationary targets, i.e. (generally) other vessels, and of your own vessel.

At first sight you would think that it would also be a useful tool for collision avoidance, and in one respect it is, because you can see the actual course and speed of other vessels. However, it has a slight problem in this regard: it can cause confusion if the vessels in the area are affected by a tidal stream.

For example, a vessel which is stationary in the water will be shown moving on the display – with the tide. The motion of vessels on the display may not be representative of their motion through the water, so you cannot make any assumption about their course and speed through the water based on their observed movement over the ground: you have to work out how the tide is affecting their movement.

True motion display: sea stabilized

The solution to this is a true motion display which is **sea stabilized**. This is a more relevant display to show vessels' motion for collision avoidance.

The picture is what you would observe from a vessel that was stationary in the water – the picture is 'fixed to the sea', and your own boat's movement through the water is shown with a moving marker.

With a sea stabilized true motion display, ground-stationary targets (land, buoys, etc) will move – in the opposite direction to the set of the tide experienced by your vessel, so the observation of land and navigation aids may become confusing. But in the open water observing other vessels, it comes into its own. As you will know from your navigation classes, all vessels on the surface of the sea in a particular locality are affected in exactly the same way by the tidal stream or current in the area. Therefore it is irrelevant in collision avoidance situations that they are all being set by the tide or current in the same direction at the same speed. A sea stabilized true motion display – if it is available – will give you a clearer picture of what target vessels around you are doing (i.e. their course and speed through the water), particularly if they change course while you are observing them.

The input data that the radar processor requires is:

■ your own vessel's heading (as for azimuth stabilised displays);

■ your own vessel's speed through the water, from an accurate log.

7

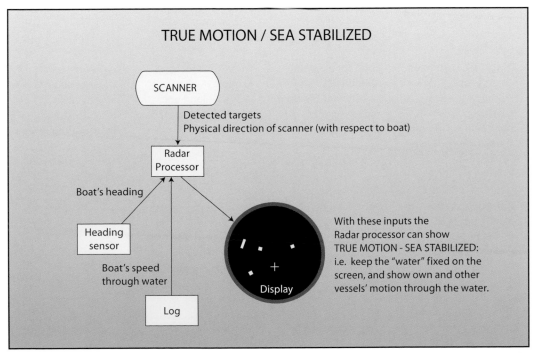

Figure 7.7 *Information block diagram showing how a True Motion Sea Stabilized display can be produced.*

Different types of display

By this time you may be confused by the different types of display and what they are good at, so the following table gives a brief summary:

	Relative motion *own vessel at centre of display*	True motion ground stabilized *display is 'fixed to ground'*	True motion sea stabilized *display 'fixed to sea', land moves in opposite direction to tidal stream*
Ship's heading-up *poor if heading not steady*	Basic display: the only one available for 'stand-alone' radar.		
Azimuth stabilized course-up *steady picture, all targets on approx same bearing to port/ starboard as observed visually*	Good for assessing collision risk (target on constant bearing) and getting CPA (by plotting target).	Good for seeing ground track of vessels – land and buoys are stationary	Good for seeing course/speed of target vessels. Other vessels on same bearing (from own boat marker) as lookout observes.
Azimuth stabilized north-up *steady picture, orientation same as chart*	Good for navigation and pilotage, and spotting collision risk (constant bearing).	Good for navigation and pilotage. Shows own and other vessels' motion over ground, oriented as chart.	

7

Signal strength

In Chapter 2, I mentioned the 'old style' and the more modern display technologies for radar. The original display is the (analogue) cathode ray tube – big, heavy, bulky and often requiring a hood for viewing. The more modern (digital) liquid crystal display is lighter, flatter and in many ways better suited to small craft.

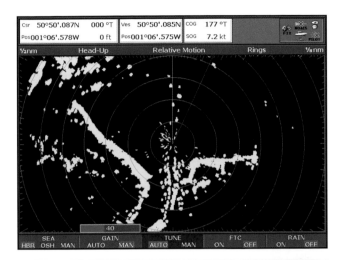

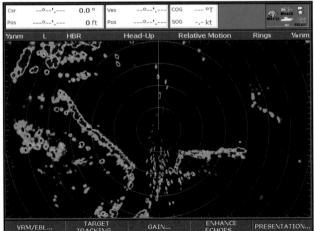

Figure 7.8 *Comparison between monochrome (top) and colour (bottom) displays. With monochrome there is no indication of signal strength; in the colour example, signal strength is coded into red, yellow, light blue, dark blue.*

However, the old display still had a couple of advantages, particularly in the early days of the new technology:

■ Because it was *not* digitized and displayed in pixels, the picture could be sharper (if properly focused). The way to improve this on modern displays is, of course, with larger displays and smaller pixels – as allowed by technology and cost.

■ The analogue display naturally showed the *strength* of the detected signal from a target – a strong signal was brighter.

The new radar on my yacht (fitted in 2005) gives no indication of signal strength. The LCD display shows all targets above a certain threshold of detected signal strength as shaded pixels on the display, and these target representations are all the same shade, regardless of the strength of signal.

More modern radars, however, do tackle this problem by providing an indication of signal strength and thereby immediately giving the observer more information via the radar picture. The most obvious indicator is colour, and with some systems the user can select the range of colours to use to represent different signal strengths. This is particularly useful when you are trying to pick out, say, a radar reflector target, like a navigation mark, amongst sea clutter (generally weak returns) in poor conditions when clutter is a particular problem.

Not all users like this all the time, so you can opt for the simple display where all targets are equal. The colours could lead to confusion, particularly if you assume that different colours mean different types of target (they don't), or that weak targets don't matter (they might), but on the whole this is a useful facility to have.

7

One manufacturer offers a simulated three-dimensional representation of the display, with the 'height' of a target representing its strength. This probably makes casual observation a bit easier, but would not be the preferred display for accurate range and bearing observation.

Dual range display

The most impressive thing about dual range display is the innovative scanner technology that goes into achieving it. However, this is of little concern to the user, so – sorry, radar engineers – I am not going to try to explain it!

What it amounts to is the ability to display two radar pictures, on different range scale settings, at the same time – as if you had two radars. This means that you don't have to keep switching from one range scale to another: you might have one set up to view distant land for navigation or to look out for fast vessels, and the other to monitor target vessels in the immediate vicinity of your own. This makes observation a great deal easier and more convenient.

The most advanced equipment allows a full choice of simultaneously displayed ranges, independently adjustable, exactly like having two independent radars. In other cases your choice may be restricted to two fairly close range settings, so that they can both use the same pulse length and pulse repetition frequency (see Chapter 6).

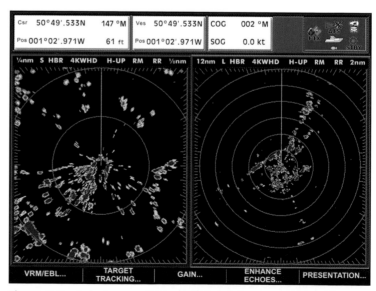

Figure 7.9 *A dual range display. Left is 1/4 nm range scale with range rings at 1/8 nm. Right is 12 nm range scale with range rings at 2 nm. This is achieved with a single radar, but it's just like having two.*

High speed update

Some radar systems offer a choice of scanner rotation speed, so that the picture can be updated more frequently. Clearly this is an advantage to the user – continuous update would be nice – but normal update speed may be enough in most circumstances. There will be technical limitations to this which may depend on range scale in use and processing speed of the radar.

Scan-to-scan correlation

Some of the signals that the radar receiver picks up and displays will be noise (i.e. not from a genuine target), and other signals will be genuine but transitory targets such as the face of a wave – sea clutter.

7

When you are looking at a radar picture, you get used to making a mental note of which targets are permanent and which are intermittent from one scan to the next. The radar can, of course, perform this function automatically, by comparing one scan with the next and not displaying intermittent targets. This is provided as an option on some systems.

The effect of this function is to clean up the picture, and it is particularly useful in clutter; but the danger is that 'naturally' intermittent, but real and important, targets may be cleaned up at the same time. Imagine a small boat with a slightly poor radar reflector bobbing around on a roughish sea – sometimes you will see it and sometimes you won't, but perhaps you would prefer to see it occasionally on your radar picture when it is detected.

This is, of course, a similar dilemma to using the clutter suppression control, and very much where the skill and practice of the operator becomes important. It also matters whether you are in a place or situation where you expect and need to see small targets. Well off-shore, in good local visibility, you might be more interested in suppressing all intermittent returns and tracking the larger targets only.

Automatic target tracking

This is a major innovation for the collision avoidance application of radar, and is mentioned in Chapter 4. This facility is dependent on two functions in the radar processor:

■ software which enables the processor to acquire, and continue to track, a particular target in the radar picture;

■ the processor's ability – when integrated with a heading sensor and GPS – to calculate the position of targets (see the section above on target latitude and longitude).

What happens is that a particular target is identified to the radar by the operator: say, by the operator moving the cursor over the target position. Alternatively, the operator can request that targets be picked up automatically, for example when they enter a **guard zone** (see section below). From then on, the radar will continue automatically to 'observe' the target as it moves – and in general it will move, both relative to your own vessel and absolutely in terms of its latitude and longitude. So the radar processor needs to keep a track of the specific target (i.e. the particular blob on the radar picture) as it changes position.

Needless to say, there could be circumstances which would make this difficult. The following factors could all be problematic:

- noise or clutter;

- other targets in the vicinity;

- the target's speed of movement;

- the size/strength/constancy of the target signal itself;

- any deficiency in the heading, speed or position data provided to the radar.

If all works well, the radar processor will be able to build up a history of the target's positions over the time it has been 'observed'. It can thereby calculate the course and speed of the target in absolute terms, and display its movement vector (perhaps a line) and its track (with, say, little dots at particular time intervals). It can also work out the Closest Position of Approach (CPA) and the time when that will occur (TCPA), assuming that both vessels – your own and the target – maintain their course and speed.

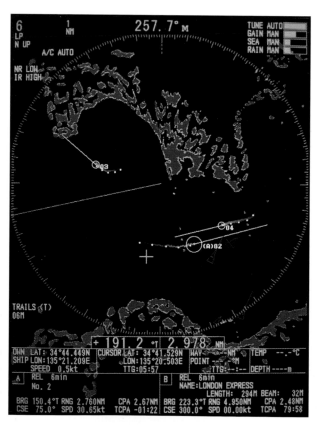

Figure 7.10 *Display showing automatic target tracking: the three white circles, with white line vectors, and white dots showing previous positions at timed intervals. Also shown are blue target wakes or echo trails, which (naturally enough) coincide with the tracks of white dots.*

The accuracy of these results depends a great deal on the accuracy of the data provided from other instruments – your own vessel's heading, speed and position. The processor needs this data to come up with the answers, and any errors in the input will result in errors in the answers. Remember, too, that CPA and TCPA will change if either vessel alters its course or speed. It may be a good idea to hold a pretty steady course and speed while using this facility in order to obtain meaningful results.

Target wakes (echo trails)

A less sophisticated, but still useful, guide to target movement is provided by the **target wakes** function (also known as **echo trails**). This simply displays where the target was on the screen on previous scans, using, for example, a less bright trail of marks for the 'wake'. It gives you a quick appreciation of how fast targets are moving on the display (the length of their trails), which is important information – and otherwise only obtainable through careful systematic observation.

If the picture is unstabilized – i.e. the basic radar display – and the heading a bit unsteady, this produces quite a mess on the screen. This is because the previous positions of the targets on the screen have more to do with the yawing of your own vessel than with actual movement of the targets! Note also that all targets are treated in this way, so land which 'moves' in relation to your vessel will also be given 'wakes'.

This function works best if you have a good azimuth-stabilized display. If it is also ground stabilized (true motion), this has the advantage that land, and stationary targets like buoys, do not generate wakes. On the other hand, if you are only observing vessels, and using the function to assist in collision avoidance, then a relative motion display will be preferable.

It is particularly useful offshore, on longer ranges, and it helps when tracking a small intermittent target so that you don't lose it. On a small boat, you don't always have the luxury of being able to concentrate exclusively on the radar; the wake can be a useful marker.

Guard zones

This is a useful and sensible function on small craft, where you don't have the luxury of a full-time radar operator. It enables you to set up an area on the screen where you want to look for targets which may be other vessels, and for the radar to automatically tell you (even wake you up) if it detects a target in that area. These are called **guard zones**. The zone selected can be very flexible in terms of its size, shape and position, and you may be able to set up more than one guard zone. (With old fashioned sets you were restricted to a simple circular band around you.) If you have automatic target tracking, you may be able to set your radar up so that it automatically starts to track any target detected in this way.

Once again, you have to remember that a target might be a genuine vessel, which itself might be small and intermittent, or it might be clutter or noise. There is a limit to how intelligent you can expect the radar processor to be in distinguishing between the two. You will probably be able to set a sensitivity level for the radar to start treating any detected blip as a genuine target, but you can see that the system is not foolproof.

For example: in rough sea, you may decide to set the threshold quite high so that you don't get annoying false alarms. If you do this, though, you might miss a weak (or intermittent) genuine target, e.g. a small vessel with a poor radar reflector. The only way to use these functions safely is to:

■ recognize their inherent limitations;

■ check with your own observations of the radar picture, and practise picking out small targets in different conditions (including adjusting gain and clutter controls);

■ keep a good visual look-out.

Watchman mode

Some radar systems have the facility to stay on standby and periodically activate themselves for a few scans. If they detect a target they give an alarm and stay active, otherwise they go back on standby. Clearly this is particularly useful in a situation where you are:

■ well away from land;

■ concerned about power usage;

■ on passage in an area where there are relatively few vessels;

■ sailing short handed.

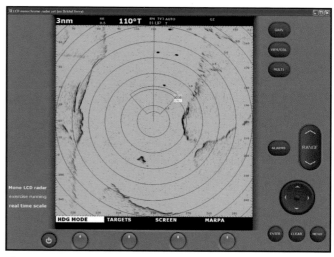

Figure 7.11 *Example of a guard zone: a wedge-shaped sector set up just ahead of the vessel's position. When a target enters the zone (and one is very close to it) the radar will give a warning.*

Conclusion

Hopefully this chapter will have explained a range of functions available on modern radars, in such a way that you can understand how they work, and what they can do for you. The advice is to talk to as many people as possible (not just salesmen!) about which functions they find useful, and when and how they use them. This will help you to choose a radar system which is right for you, and you won't spend extra money on unnecessary features.

Ultimately, radar is an instrument that can be used in many different ways – you have to develop your own style and preferences through experience.

8 Choosing and installing radar

It is not the purpose of this book to provide you with a full Do It Yourself guide on installing radar on your boat. Many readers will have bought or chartered vessels with the instruments already installed – others will prefer to rely on experienced professionals. Some may wish to do the work themselves, and their primary source of guidance will be the manufacturers' installation guides.

What this chapter does is to discuss and explain the issues surrounding the selection and installation of radars, so that you can understand the choices to be made and the technical issues to be considered, specifically with regard to radar – whether or not you want to get involved with the physical work!

A key factor in selecting a radar is its functionality compared to the alternatives, and what suits your purposes. The functionality of modern radars is discussed in other chapters (mainly Chapter 7), so those choices and issues are not repeated here.

Safety

These days we are bombarded with safety advice as soon as we open any user guide, be it for a washing machine or a motor car, and who are we to disagree with this. Certainly radar is no exception, and the warnings are very relevant. You are advised to follow the manufacturer's guidance and advice, seeking qualified assistance if necessary. This book does not provide that!

There are three radar-specific health warnings in particular, which all users should be aware of:

■ **Electromagnetic energy.** The scanner transmits microwaves at high power, and exposure to this should be avoided. This influences the choice of scanner position (it should be installed in a suitable location to avoid exposure to people), and also means that the equipment should be switched off if anyone is working near the scanner. (The effect of microwaves should be familiar to anyone who uses a microwave oven.)

■ **High Voltage.** Manufacturers advise that both the display units and the scanner units contain high voltage, and should not be opened by users in an attempt to fix or service them. This is particularly worth heeding in an environment which might be wet with salt water.

- **Navigation aid.** Although not specifically an installation issue, I have repeatedly stated in this book that you should not place too much reliance on radar, because of its inherent limitations and because it may or may not be:

  - installed correctly

  - working properly; or

  - being used effectively.

Radar power

The choice of **radar power** will probably be 2 or 4 kW for smaller scanner units, with higher powers available – say, 10 or 12 kW for larger scanners.

Power is obviously necessary to get good returns from poor or distant targets, so to that extent, the more the better. But remember that, for most small craft, the range will be determined more by the height of the scanner above sea level. There is no point in installing a high powered radar with range scale of 48 miles in a position where the scanner can only 'see' 15 miles to the radar horizon.

The trade-offs that you may need to consider when selecting the best power for your vessel are:

- cost (obviously);

- scanner size and weight (see below);

- power needed to run the equipment (particularly on sailing yachts).

The advice is to decide what sort of maximum range you realistically need and can achieve – which probably depends mainly on scanner positioning – and then buy a set with sufficient power to give reasonable performance at this range.

Scanner size

Scanner size determines another aspect of radar performance: beam width. The larger the scanner, the narrower the beam width, and therefore the better the angular discrimination of targets.

Typically, an 18-inch dome will achieve a beam width of about 5 degrees, and a 24-inch dome will give a slight improvement: around 4 degrees. If you want to do better than this, you will need a larger and heavier scanner, and probably need to look at higher power so that the radar can achieve adequate performance with the narrower beam. There would be no point to this unless you also installed a system with a fairly high definition display, so you are moving up the scale on pretty well all fronts.

Generally the manufacturers will provide 'packages' – power, scanner size, display size etc – which are sensible combinations for the price being paid.

Scanner weight

Apart from the obvious question of the structural support, scanner weight may affect boat stability, particularly if the scanner is installed high up the mast.

In simple terms, the lower a boat's centre of gravity, the more stable it is, and boat designers take pains to achieve this – for example, by installing ballast weight in the keel. By placing a heavy piece of equipment high above the waterline you counteract these efforts, and the effect is proportional to:

■ the weight of the scanner; and

■ the height at which it is installed.

It is sensible to seek advice on this, particularly if you want to install a fairly large radar on a small yacht or motor boat. Most vessels should still have adequate stability if you are sensible about your choice of equipment and follow the vessel manufacturer's guidelines.

Small radar domes typically weigh around 12kg, but larger scanners can weigh 35kg or more, and so are really suited to larger vessels.

Scanner position

The following factors will affect the choice of scanner position:

■ the higher the scanner, the greater the radar horizon range (the equation and examples are given in Chapter 2);

■ its size and weight will constrain where it can be installed: the top of the mast generally isn't an option, and it needs to be clear of other equipment, sails and ropes;

■ greater height will have a greater adverse effect on stability;

■ it needs to be positioned clear of structures and equipment that will cause blind spots or radar reflections;

■ for safety, it should be positioned where people on board will not be affected by the electromagnetic radiation.

8

Blind spots and own ship radar reflections are generally problems for larger ships with their more complex superstructures.

In practice there will be relatively few available positions on most small craft – e.g. up the mast, on top of the wheelhouse, or on a separate 'A' frame or stub mast.

Remember also that the scanner needs to be level, so that the beam will sweep the surface of the water all round the vessel. Some craft, particularly power vessels designed to operate on the plane, will sit at a slightly different angle planing to when they are in displacement mode or at rest. If the beam is angled too far upwards when planing, targets close in front of the boat may be missed.

It pays to look at installations on similar craft to get some ideas: then seek advice and decide, based on the above factors.

Electrical supply and cables

The main factor to note about the electrical supply is that the power requirement is comparatively high. Therefore on a low voltage (12V or 24V), it is necessary to use reasonably heavy cable, particularly on long runs, to avoid loss of voltage. This applies to the main power supply, and also to the run between the display and the scanner – which may be relatively long if the scanner is installed up the mast, for example.

Some units have 'booster boxes' to increase the voltage to the scanner. Essentially these are transformers, turning low voltage/high current into high voltage/low current, which results in reduced losses in the cable run – so the cable size becomes less critical for a long run.

New radars come supplied with the correct cables, as well as detailed instructions for installing and extending them if needed. These instructions should be followed carefully.

Cables containing multiple small wires can be damaged if they are bent through too tight a radius, so take care if you are feeding them through confined spaces or installing the display in a small console.

Integration with other instruments

The actual process of connecting with other instruments can be quite straightforward with the standards **(SeaTalk, NMEA)** that allow different instruments to exchange data and interact with each other. It could be more problematic if you try to integrate with older, less compatible equipment.

Perhaps the more important point here is the selection of good equipment to integrate with. The standards will allow you to integrate with virtually anything, including, for example, heading sensors which are inadequate for the task of providing good azimuth stabilized displays. In fact, the equipment is so flexible that it will probably take 'direction' information from anywhere it can find it.

If you don't have a heading sensor, it is quite capable of using the GPS course over ground (COG) instead, which will 'work' in the sense that it will get the data and give you a course-up or north-up display, but will work *badly* because COG has very little relation to instantaneous ship's heading (or none if the boat is stationary).

Again, advice from manufacturers, dealers and specialist installers is important. This may mean that a manufacturer will advise you to integrate with their own range of equipment rather than mixing with other makes of equipment, and you will have to decide if this is the best course of action for you.

Positioning of display

Nowadays you have quite a lot of choice in where to position the display, because LCD displays are far smaller than the previous CRT displays (the same probably applies to your new television at home).

Given that the same display may well be a chart plotter, this may influence your choice. It is heavily dependent on your boat, and how you use your boat: personally, on a sailing yacht, I choose to position the display at the navigation station below deck, but angled so that I can see it from the companionway (watch out for inconvenient light reflections though). Others will prefer more than one display, and essentially we are spoilt for choice here.

One other point: it is instinctively easier to use the display in heading-up or course-up mode if the display is oriented on the boat so that you view it facing forwards. Much of what you do involves comparing the display to the view from the boat, and this orientation seems to help considerably.

Conclusion

Careful thought is important for the installation of any equipment on a boat, and making these choices on your own boat is a considerable source of satisfaction.

There is quite a lot to be considered in the case of radar. It is a big investment both financially and in terms of the 'equipment budget' for your vessel, given that there is a limit to the amount of gear that you will want to buy, install, power, maintain, train on and use. So it is worth understanding the issues clearly when deciding how to proceed.

Good luck with your installation!

Glossary/Index

The following table combines a **glossary** of technical terms which are used in the book, and an **index** of topics which are covered, with a reference in each case to where they are discussed fully in the text.

Term/topic	Explanation	Pp
A		
active radar reflector	transponder device which transmits a radar signal when scanned	37, 38
automatic identification system (AIS)	system for sending/receiving information about ships, to aid collision avoidance	35, 49-51, 89
automatic radar plotting aid (ARPA)	system on ships for tracking radar targets	43
automatic target tracking	process of tracking radar targets to aid collision avoidance	43, 49, 95-97
azimuth stabilized	radar picture which is 'fixed' to a particular direction, usually north or ship's intended course (see: north-up, course-up)	17, 18, 83, 84
B		
backlight	setting on LCD display	26
beam	see: radar beam	
beam width	angular width of radar beam (see: horizontal beam width, vertical beam width)	17, 72-74
bearing discrimination	accuracy of bearing (direction) measurement of a target: ability to distinguish two targets at the same range close together in bearing	72, 73
brilliance	for CRT displays: brightness at which the radar picture is displayed	26, 27
C		
cables		102
cathode ray tube (CRT)	older technology for radar display (similar to older TVs)	23, 93
chart: comparison to radar		13, 14, 55-57, 88
clearing line	a line drawn on a chart which has safe water on one side and dangers on the other	59, 65, 66
closest point of approach (CPA)	closest distance at which a target ship will pass, calculated automatically or from plotting	43-46, 95-97
clutter	unwanted radar returns, e.g. from sea or precipitation	22, 23, 29-31
collision risk	when two vessels are in danger of colliding (see: International Regulations for Preventing Collisions at Sea: IRPCS)	39-43

Term/topic	Explanation	Pp

P

Term/topic	Explanation	Pp
passive radar reflector	radar reflector which simply reflects incoming radar signals (see also: active radar reflector)	37
pilotage	navigation close to dangers, e.g. out of or into harbour	54-68
plan position indicator (PPI)	type of radar display used on boats, showing a plan view of targets around the vessel	12, 13, 28
plotting	see: radar plotting	
position fix	process of determining ship's position	62, 63
position line	line on a chart, on which the ship's position lies	58, 59-61
positioning of display		103
power	see: radar power	
precipitation clutter	see: rain clutter	22, 23, 30, 31
pulse	short burst of transmitted microwaves	15, 16
pulse length	duration of radar pulses transmitted	70-72
pulse repetition frequency (PRF)	number of radar pulses transmitted per second	70

R

Term/topic	Explanation	Pp
racon (radar beacon)	transponder used to provide radar identification of navigation marks	63
radar	acronym for RAdio Direction And Ranging	
radar assisted collision	collision which was probably caused by misuse of radar	40
radar beam	narrow beam of transmitted microwaves	11, 16, 17
radar clearing line	clearing line determined by radar observation	59, 65, 66
radar horizon	point at which the sea surface becomes invisible to radar because of the earth's curvature	19, 20, 78, 79
radar plotting	process of systematically recording target positions for collision avoidance	43-48
radar position fix	process of determining ship's position from radar position lines only	62, 63
radar position line	position line determined by radar observation	59-62
radar power	power of transmitted signal	21, 69, 70, 100
radar range	maximum range at which targets can be detected (due to curvature of the earth, height of target, radar power etc.)	19, 20, 69, 70, 100
radar reflector	device for enhancing the radar visibility of a small vessel or buoy	21, 37, 38
rain clutter	unwanted radar returns from rain or other precipitation	22, 23, 30, 31
range discrimination	accuracy of range measurement of target: ability to distinguish two targets on the same bearing close together in range	71, 72

Term/topic	Explanation	Pp
range rings	concentric circles on display showing set intervals of range	26, 28
range scale	selection of scale for the display, defined by maximum range of targets shown on the display	26, 38
raster	technology for digital display of pictures (as opposed to analogue)	23
rate gyro	device which can sense rate of change of direction	85
refraction: effect on radar range		78, 79
relative motion	display where own ship is at the centre, and movement of targets is relative to own ship position	39, 90, 92
return	*see: echo return, target*	

S

Term/topic	Explanation	Pp
safety warnings		99, 100
satellite compass	*see: GPS compass*	85
S-band	one of two frequency bands used by marine radars *(see also: X-band)*	69
scanner position		101, 102
scanner size		100, 101
scanner weight		101
scan-to-scan correlation	process of automatically comparing targets detected on successive scans of the radar	95
sea clutter	unwanted radar returns from the sea	22, 29, 30
sea stabilized	display which is 'fixed' to the surface of the sea, showing own and target ships' movement through the water	90-92
search and rescue transponder (SART)	transponder device which causes an emergency signal to be displayed on a radar screen	64
Seatalk	propriatory specification for communication between different marine electronic devices	102, 103
ship's heading marker	line on the display showing own ship's heading	28, 84
ship's heading sensors	*see: heading sensors*	85
ship's heading-up	basic form of radar display which is not azimuth stabilized: ship's heading is the '12 o'clock' direction	16, 28, 92
side lobe	secondary radar beams either side of the main beam	74-76
signal strength: by type of target	*see: target strength (by type)*	21, 22
signal strength: indication on display	display where strong and weak target signals are distinguished in some way	93, 94
signal strength: relation to range		69, 70

Term/topic	Explanation	Pp
stability: effect on		101
stabilization	*see: azimuth, ground and sea stabilization*	
standby	mode where radar is warmed up and ready to transmit, but not transmitting	25, 26, 98

T

Term/topic	Explanation	Pp
target	object detected and displayed by radar	12
target expansion	facility to make targets larger and more visible on the display	70, 72
target latitude/longitude	facility to determine the latitude/longitude position of targets	86-87
target strength (by type)		21, 22
target wakes	facility to display track of previous target positions on the display	97
time to closest point of approach (TCPA)	estimate of when a target will arrive at its closest point to own vessel, calculated automatically or from plotting	43-46, 95-97
transmit	control to start radar operating *(see also: standby)*	26
transponder	device which transmits a response on detecting a radar signal *(see also: racon; active radar reflector; SART)*	37, 38, 63, 64
true motion	another term for ground or sea stabilized displays, showing the actual motion of own and target ships over the ground or through the water	55, 90-92
tuning	process of fine-tuning the radar receiver to the exact frequency of the transmitter	28

V

Term/topic	Explanation	Pp
variable range marker (VRM)	adjustable circular line on the radar display showing a particular range from the ship	31, 32, 59, 60, 66
vertical beam width	angular width of radar beam in the vertical direction	74

W

Term/topic	Explanation	Pp
watchman mode	facility for radar to automatically switch itself on to check for targets	98

X

Term/topic	Explanation	Pp
X-band	one of two frequency bands used by marine radars *(see also: S-band)*	69

Z

Term/topic	Explanation	Pp
zoom-in	expand a particular part of the radar picture	82

References for further study

Study commissioned as a result of the loss of the yacht Ouzo:

Steve Luke, *Performance Investigation of Marine Radar Reflectors on the Market*. QinetiQ Ltd, March 2007. Can be accessed from www.maib.gov.uk (see the investigation report on Ouzo, 2007).

Recommended for more detailed study:

Alan Bole, Bill Dineley and Allan Wall, *Radar and ARPA Manual*, Elsevier Butterworth Heinemann, 2005 (ISBN: 0 7506 6434 7). First published 1990, second edition 2005, reprinted 2006. Various reprints and paperback edition are available.

See also:

RYA Shorebased Course: *Radar Course*. This is a one-day course offered by approved RYA Training Centres (see www.rya.org.uk for contacts and further details of the course).

Raymarine offer one-day radar courses by experienced radar instructors

You can experience realistic hands-on operation of small-craft radar, using a Simulator program running on your PC. You control your own vessel, and the coastline, target vessels, buoys, weather etc. are all variable, for unlimited experience.

LightMaster Software is the UK's leading producer of training software for marine leisure. LightMaster training and simulation software has been adopted as the standard for use in VHF and Radar courses.

Other products comprise animated training programs covering VHF Radio, Collision Regulations, Buoys, and GPS Navigation. Single-user versions of each program can be purchased at low cost to provide valuable preparation for courses and realistic ongoing revision, without the risks such as false alerts or collisions which might result from practising on real equipment or vessels.

All LightMaster software runs on standard PCs with Windows(tm) and is available from: www.lightmaster.co.uk

Acknowledgements

My thanks to: Bill Anderson, Pat Manley and Peter Weirman for reviewing the manuscript and diagrams and for their invaluable comments, based on very much greater experience than my own.

Thanks to Raymarine for their permission to use their copyrighted photographs, screenshots and data about their radar systems (12, 13, 25, 27, 31, 50, 88, 93, 94), and to Neil Millerchip and Roger Arnott for their assistance.

Thanks to Furuno UK for their photographs and screenshots (12, 30, 31, 82, 96) and to Bruce Hardy for his assistance.

Thanks to Ivor Hathaway for supplying his photograph (page 36).

Thanks to Lightmaster for creating screenshots.

The istock images on page 7 (©istock.com/Daniel Dupuis) and page 38 (©istock.com/choppy) used with permission.

Thanks to Tideland Signal Ltd for supplying racon image (page 63).

Thanks to McMurdo for supplying SART image (page 64).